Rapid Assessment Program

RAP Bulletin of Biological Assessment 44

A Rapid Biological Assessment of North Lorma, Gola and Grebo National Forests, Liberia

Peter Hoke, Ron Demey and Alex Peal
(Editors)

Center for Applied Biodiversity Science (CABS)

Conservation International

Conservation International – Liberia

Forestry Development Authority (FDA)

Society for the Conservation of Nature (SCNL)

University of Liberia

United Nations Mission in Liberia

United States Department of State

The *RAP Bulletin of Biological Assessment* is published by:
Conservation International
Center for Applied Biodiversity Science
2011 Crystal Drive, Suite 500
Arlington, VA 22202
USA

703-341-2400 telephone
703-979-0953 fax
www.conservation.org
www.biodiversityscience.org

Editors: Peter Hoke, Ron Demey and Alex Peal
Design/production: Kim Meek
Map: Mark Denil

***RAP Bulletin of Biological Assessment* Series Editors:**
Terrestrial and AquaRAP: Leeanne E. Alonso
Marine RAP: Sheila A. McKenna

Conservation International is a private, non-profit organization exempt from federal income tax under section 501 c(3) of the Internal Revenue Code.

ISBN: 978-1-934151-01-3

U.S. Library of Congress Catalog Card Number: 2007923847

DOI: 10.1896/ci.cabs.2007.rap44.liberia

RAP Bulletin of Biological Assessment was formerly *RAP Working Papers.* Numbers 1–13 of this series were published under the previous title.

Suggested citation: Hoke, P., R. Demey and A. Peal (eds.). 2007. A rapid biological assessment of North Lorma, Gola and Grebo National Forests, Liberia. RAP Bulletin of Biological Assessment 44. Conservation International, Arlington, VA, USA.

Table of Contents

Participants and Authors

Abdulai Barrie (large mammals)
Center for Biodiversity Research
7 Duke Street
Freetown, SIERRA LEONE
Email. ahbarrie@yahoo.com

Amandu K. Daniels (plants)
Forestry Development Authority
P.O. Box 10-3010
1000 Monrovia 10, LIBERIA

Ron Demey (birds, editor)
Van der Heimstraat 52
2582 SB Den Haag, NETHERLANDS
Email. rondemey@compuserve.com

Klaas-Douwe B. Dijkstra (invertebrates)
Gortestraat 11
2311 MS Leiden, NETHERLANDS
Email. dijkstra@nnm.nl

Jakob Fahr (contributing author)
Department of Experimental Ecology (Bio III)
University of Ulm
Albert-Einstein Allee 11
D - 89069 Ulm, GERMANY
Email. jakob.fahr@uni-ulm.de

Theo Freeman (logistics coordinator)
Forestry Development Authority
P.O. Box 10-3010
1000 Monrovia 10, LIBERIA

Joel Gamys (large mammals)
Conservation International – Liberia (CI)
Smythe Road, Old Road
Sinkor, Monrovia, LIBERIA
Email. gamysjoel@yahoo.fr

Moses G. Gorpudolo (plants)
University of Liberia
Monrovia, LIBERIA

Annika Hillers (reptiles and amphibians)
Institute for Biodiversity and Ecosystem Dynamics (IBED)
University of Amsterdam
Kruislaan 318
1098 SM Amsterdam, NETHERLANDS
Email. ahillers@yahoo.com

Peter Hoke (logistics coordinator, editor)
Rapid Assessment Program
Conservation International
2011 Crystal Drive, Suite 500
Arlington, Virginia, USA
Email. phoke@conservation.org

Carel Jongkind (plants)
Wageningen University
Tarthorst 145
6708 HG Wageningen, NETHERLANDS
Email. Carel.Jongkind@wur.nl

John Konie (plants)
University of Liberia
Monrovia, LIBERIA

Aaron N. Kota, Sr. (large mammals)
Forestry Development Authority
P.O. Box 10-3010
1000 Monrovia 10, LIBERIA

Mawolo Kpewor (small mammals)
University of Liberia
Monrovia, LIBERIA

Roger Luke (large mammals)
Forestry Development Authority
P.O. Box 10-3010
1000 Monrovia 10, LIBERIA

Miaway Luo (large mammals)
Forestry Development Authority
P.O. Box 10-3010
1000 Monrovia 10, LIBERIA

Ara Monadjem (small mammals)
University of Swaziland
UNISWA, Private Bag 4
Kwaluseni, SWAZILAND
Email. ara@uniswacc.uniswa.sz

Flomo Molubah (birds)
Society for the Conservation of Nature in Liberia
Monrovia Zoo, Lakpazee
P.O. Box 2628
Monrovia, LIBERIA

Joshua Quawah (small mammals)
Forestry Development Authority
P.O. Box 10-3010
1000 Monrovia 10, LIBERIA

Mark-Oliver Rödel (contributing author)
Department of Animal Ecology and Tropical Biology
Biocenter
Am Hubland, D-97074 Würzburg, GERMANY
Email. roedel@biozentrum.uni-wuerzburg.de

Richard Sambolah (large mammals)
Fauna & Flora International
Monrovia, LIBERIA

Evangeline Swope (birds)
Forestry Development Authority
P.O. Box 10-3010
1000 Monrovia 10, LIBERIA

Sormongar Zwuen (large mammals)
Forestry Development Authority
P.O. Box 10-3010
1000 Monrovia 10, LIBERIA

FIELD SUPPORT

Jerry Brown (logistics)
Conservation International – Liberia
Smythe Road, Old Road
Sinkor, Monrovia, LIBERIA

Henry Gardner (driver)
Society for the Conservation of Nature in Liberia
Monrovia Zoo, Lakpazee
P.O. Box 2628
Monrovia, LIBERIA

Morris S. Kamara (driver)
Environmental Foundation for Africa / Fauna & Flora International
Monrovia, LIBERIA

Charles Kollie (cook)
Mamba Point Hotel
Monrovia, LIBERIA

Nyumah Mensoh (logistics)
Conservation International – Liberia
Smythe Road, Old Road
Sinkor, Monrovia, LIBERIA

PROJECT SUPPORT

Amos Andrews (communication)
Conservation International – Liberia
Smythe Road, Old Road
Sinkor, Monrovia, LIBERIA

W. Tyler Christie (coordinator)
Conservation International – Liberia
Smythe Road, Old Road
Sinkor, Monrovia, LIBERIA

Alex Peal (editor)
Conservation International – Liberia
Smythe Road, Old Road
Sinkor, Monrovia, LIBERIA

Zinnah Sackie (finances)
Conservation International – Liberia
Smythe Road, Old Road
Sinkor, Monrovia, LIBERIA
Email. zsackie@conservation.org

Nathaniel Walker (logistics)
Conservation International – Liberia
Smythe Road, Old Road
Sinkor, Monrovia, LIBERIA
Email. nwalker@conservation.org

Organizational Profiles

CENTER FOR APPLIED BIODIVERSITY SCIENCE (CABS)

The Center for Applied Biodiversity Science (CABS), the scientific hub of Conservation International, works to link science and action to guide the conservation of nature worldwide.

Conservation initiatives have garnered significant political support in the last quarter century, as is evident in the international consensus around instruments such as the 1992 Convention on Biological Diversity. Political and economic support alone, however, is not enough to preserve the Earth's dwindling biodiversity. The conservation community can only be effective if it is equipped with clear goals, objectives, and strategies grounded in reliable and verifiable scientific research. There is still much to learn about the Earth's natural diversity, its role in ecosystem function and related services, and the most effective ways to preserve it.

Scientists at CABS work to fill these knowledge gaps. Founded in 1999 with generous support from the Gordon and Betty Moore Foundation, CABS brings together a staff of more than 70 research scientists who are highly respected in their fields and dedicated to saving our biodiversity.

CONSERVATION INTERNATIONAL

Conservation International (CI) is an international, non-profit organization based in Arlington, VA. CI's mission is to conserve the Earth's living natural heritage, our global biodiversity, and to demonstrate that human societies are able to live harmoniously with nature.

Conservation International
2011 Crystal Drive, Suite 500
Arlington, VA 22202 USA
Tel. 1-703-341-2400
Fax. 1-703-553-0654
Web. www.conservation.org
www.biodiversityscience.org

CONSERVATION INTERNATIONAL–LIBERIA

Conservation International–Liberia has been working with the Government of Liberia and civil society organizations since 2002 to achieve its conservation goals. These include: working with the Government of Liberia to create a network of protected areas covering 1.5 million hectares of the remaining forest cover, strengthening and improving protected area and wildlife management; increasing awareness and public participation; enhancing livelihoods by fostering sustainable biodiversity use; and promoting good governance of Liberia's forest resources. CI–Liberia implements its strategy to conserve Liberia's biodiversity by creating partnerships, conducting scientific research, and improving human welfare.

Conservation International–Liberia
Smythe Road, Old Road
Sinkor, Monrovia
LIBERIA
Tel. 231-6-511-138

FORESTRY DEVELOPMENT AUTHORITY (FDA)

A 1976 Act of Legislature created the Forestry Development Authority (FDA) with responsibilities to effectively conserve and sustainably manage the forest resources of Liberia for its entire people. Among the primary objectives of the Authority are: (a) Establish a permanent forest estate made up of reserved areas upon which scientific forestry will be practiced; (b) Conduct essential research in forest conservation and pattern action programs upon the results of such research; (c) Give training in the practice of forestry; offer technical assistance to all those engaged in forestry activities; and spread knowledge of forestry and the acceptance of conservation of natural resources throughout the country;

(d) Conserve recreational and wildlife resources of the country concurrently with the development of forestry programs.

Forestry Development Authority
P.O. Box 10-3010
1000 Monrovia 10
LIBERIA

SOCIETY FOR THE CONSERVATION OF NATURE IN LIBERIA (SCNL)

SCNL is a non-governmental, non-profit conservation organization in Liberia. The organization was created in 1986 and has since worked to promote programs that influence the wise use of biological resources in Liberia, Africa and the world. Its mission is to manage and conserve biodiversity; increase public awareness of the importance of sustainable utilization of natural resources; improve livelihoods to alleviate poverty and work toward maintaining the integrity of the natural environment in Liberia.

Society for the Conservation of Nature in Liberia
Monrovia Zoo, Lakpazee
P.O. Box 2628
Monrovia
LIBERIA
Tel. 231-6-572377/ 231-6-512506/ 231-6-557441

UNITED NATIONS MISSION IN LIBERIA (UNMIL)

The United Nations Mission in Liberia (UNMIL) was established by Security Council resolution 1509 (2003) of 19 September 2003 to support the implementation of the ceasefire agreement and the peace process; protect United Nations staff, facilities and civilians; support humanitarian and human rights activities; as well as assist in national security reform, including national police training and formation of a new, restructured military. UNMIL also has a mandate "to assist the transitional government in restoring proper administration of natural resources." Under such mandate, UNMIL has been supporting Government's various natural resources management programme/projects, including the provision of assistance in forest resource management and wildlife protection.

UNMIL Headquarters
Pan African Plaza
Tubman Boulevard, 1st Street
Monrovia, Liberia
Tel. 231-6-566566

UNITED STATES DEPARTMENT OF STATE

The United States Department of State works to create a more secure, democratic, and prosperous world for the benefit of the American people and the international community.

U.S. Department of State
2201 C Street NW
Washington, DC 20520 USA
Tel. 202-647-4000

UNIVERSITY OF LIBERIA

The University of Liberia (UL), formerly Liberia College, was created in 1872. A president appointed by the President of the Republic, with oversight by a Board of Trustees, heads the University of Liberia. A Council and a Faculty Senate is responsible for the day-to-day management of the institution.

UL has three graduate schools – Regional Planning, Educational Administration and Supervision and the Ibrahim B. Bagagida School of International Studies. The UL also runs three specialized programs, the Louis Arthur Grimes School of Law, The A. M. Doglioti College of Medicine and the School of Pharmacy.

There are five undergraduate programs at the University of Liberia:

1. The Liberia College (College of Social Science and Humanities)
2. The William V. S. Tubman Teacher College
3. The William R. Tolbert, Jr. College of Agriculture and Forestry
4. The College of Business and Public Administration
5. The T. J. R. Faulkner College of Science and Technology

University of Liberia
P.O. Box 9020
Monrovia
LIBERIA
Tel. 231-6-422-304

Acknowledgments

Conservation International–Liberia and the entire RAP team would like to thank the United States Department of State for funding this expedition. We also thank the Government of Liberia and the Forestry Development Authority (FDA), and in particular the Managing Director, D. Eugene Wilson, for their support and guidance.

We are extremely grateful for the logistical support that was provided by the United Nations Mission in Liberia (UNMIL); without it this expedition would have been impossible. During our meeting with Alan Doss, Special Representative to the Secretary General, it was apparent that he recognized the importance of Liberia's flora and fauna. He is well served by his advisors, Webby Bonali, Senior Advisor/Environment and Hiroko Mosko, Advisor/Environment. Both were instrumental in this project and in coordinating our movement throughout Liberia. Special thanks to Hiroko for her tireless efforts making sure that we made it safely to each of our destinations. At our sites in North Lorma and Grebo National Forests, Lt. Col. Syed Wajid Raza and his men of PAK BAT IV in Voinjama provided us with a safe environment to do our work. In Grebo, the UN staff at Fishtown and the Ethiopian Army assisted us. We are also grateful to the UN helicopter pilots and crew that safely and swiftly transported us to each of our survey sites.

We appreciate the generous hospitality of each of the communities that we stayed in prior to entering the forest. Chief Flomo Zuba of Luyema, Chief Seku Kamara of SLC and Superintendent Christian Chea of Jalipo welcomed us and facilitated the hiring of guides and assistants. We would like to thank the numerous guides, assistants and cooks that helped us during our visits.

The participants would like to thank all of the personnel at Conservation International–Liberia, in particular Tyler Christie, Zinnah Sackie, Nat Walker, Amos Andrews, Jerry Brown, and Nyumeh Mensoh for organizing this survey in Liberia.

We thank Professor Blaydon of the University of Liberia for allowing the team use of botanical equipment, Fauna & Flora International–Liberia for the use of a vehicle and the University of Liberia for hosting the closing event where we presented our preliminary results. We would like to recognize our colleagues in CI–Ghana for assisting with Ara Monadjem's travel through Accra for this survey. We also owe thanks to Mark Denil of CI's Conservation Mapping Program and Jennifer McCullough for her comments on the report.

The small mammal team would like to thank Mawolo Kpewor and Joshua Quawah for assistance in the field. The identification of shrews and murids by Rainer Hutterer (ZFMK) is much appreciated. Fritz Dieterlen (SMNS) and Wulf Gatter, Lenningen, kindly made available bat specimens collected in Liberia by the latter. Suzanne B. McLaren (CM) sent bat specimens on loan and the late Charles O. Handley, Jr. provided access to the mammal collections at the USNM. Jan Decher, University of Vermont, commented on a draft version of the manuscript. Analysis and publication of the data is part of the BIOLOG-program of the German Ministry of Education and Science (BMBF; project W09 BIOTA-West, 01 LC 0411).

Lastly, this was a logistically challenging RAP and we appreciate the patience of the scientists and assistants that participated. Despite difficult conditions at times, everyone kept a good attitude and worked hard to document Liberia's amazing biodiversity.

Report at a Glance

A RAPID BIOLOGICAL ASSESSMENT OF NORTH LORMA, GOLA AND GREBO NATIONAL FORESTS, LIBERIA

Expedition Dates
13 November – 11 December 2005

Area Description
North Lorma National Forest consists of seasonal moist evergreen and semi-deciduous forest and some open riverine forest habitat. It is situated in northwestern Liberia near the border with Guinea and lies between the Wologizi and the Wonegizi Mountains. Many smaller streams were present within a slightly hilly landscape. Further from the river but at a higher elevation the vegetation quickly changed to lower forest with large scattered trees. This site had the least amount of disturbance with the presence of an old overgrown logging road the only noticeable sign.

Gola National Forest is also seasonal moist evergreen and semi-deciduous forest and is situated between the Gola Strict Nature Reserve in Sierra Leone and Kpelle National Forest in Liberia. The terrain has steep slopes with small, rocky streams and the vegetation was dense in most areas although large lianas were present. Illegal small-scale diamond mining was observed just inside the forest.

Grebo National Forest is a wet evergreen forest situated in the southeast of the country and is contiguous with the Forêt Classée du Cavally in Côte d'Ivoire. It consists of mature secondary forest that is open with isolated huge trees. Aquatic sites within the area were medium sandy streams with a few stones and rocks as well as large ponds. Logging occurred in this area roughly twenty years ago.

Reason for the Expedition
This survey was part of the larger Liberia Forest Initiative (LFI), an initiative supporting efforts to rehabilitate and reform the forest sector in Liberia and harmonize activities associated with these efforts. In addition to collecting data on the sites to strengthen and expand conservation efforts in Liberia, the RAP survey worked to build scientific capacity within the FDA, University of Liberia and local NGOs. Efforts were also made to increase the Liberian public's awareness of their rich flora and fauna.

Major Results

	All RAP sites in this survey	North Lorma	Gola	Grebo
Number of species recorded	969	526	486	520
Species of conservation concern	60	29	27	41
New species discovered	6	—	5	4
New records for Liberia	18	7	7	3
Species endemic to Upper Guinea	147	61	81	70

Number of Species Recorded:

Plants	548 species
Dragonflies & damselflies	93 species
Amphibians	At least 40 species
Reptiles	17 species
Birds	211 species
Bats	22 species
Small terrestrial mammals	9 species
Large mammals	29 species

New Species Discovered:

Plants (3)
- *Drypetes* sp. nov.
- *Leptoderris* sp. nov.
- *Rhaphiostylis* sp. nov.

Dragonflies and damselflies (1)
- *Eleuthemis* sp. nov.

Amphibians (1?)
- possibly *Phrynobatrachus* cf. *annulatus*

Bats (1?)
- possibly *Neoromicia* aff. *grandidieri*

New Records for Liberia:

Plants (3)
- *Elytraria ivorensis*
- *Gardenia nitida*
- *Zanthoxylum psammophilum*

Dragonflies and damselflies (7)
- *Nesciothemis minor*
- *Palpopleura deceptor*
- *Palpopleura portia*
- *Paragomphus nigroviridis*
- *Tetrathemis polleni*
- *Tramea limbata*
- *Trithemis monardi*

Amphibians (5)
- *Afrixalus nigeriensis*
- *Astylosternus occidentalis*
- *Bufo superciliaris*
- *Chiromantis rufescens*
- *Phrynobatrachus villiersi*

Bats (3)
- *Neoromicia* aff. *grandidieri*
- *Neoromicia guineensis*
- *Rhinolophus landeri*

Species of Conservation Concern:

Dragonflies and damselflies (2)
- *Sapho fumosa* (NT)
- *Trithemis africana* (NT)

Amphibians (17)
- *Amnirana occidentalis* (EN)
- *Phrynobatrachus annulatus* (EN)
- *Phrynobatrachus* cf. *annulatus* (EN)
- *Conraua alleni* (VU)
- *Phrynobatrachus villiersi* (VU)
- *Afrixalus nigeriensis* (NT)
- *Bufo togoensis* (NT)
- *Hyperolius chlorosteus* (NT)
- *Leptopelis macrotis* (NT)
- *Leptopelis occidentalis* (NT)
- *Petropedetes natator* (NT)
- *Phrynobatrachus alleni* (NT)
- *Phrynobatrachus guineensis* (NT)
- *Phrynobatrachus liberiensis* (NT)
- *Phrynobatrachus phyllophilus* (NT)
- *Ptychadena superciliaris* (NT)
- *Bufo superciliaris* (CITES I)

Reptiles (5)
- *Osteolaemus tetraspis* (CITES I)
- *Kinixys eros* (CITES II)
- *Kinixys homeana* (CITES II)
- *Python sebae* (CITES II)
- *Varanus ornatus* (CITES II)

Birds (14)
- *Malimbus ballmanni* (EN)
- *Agelastes meleagrides* (VU)
- *Bleda eximius* (VU)
- *Criniger olivaceus* (VU)
- *Lobotos lobatus* (VU)
- *Melaenornis annamarulae* (VU)
- *Picathartes gymnocephalus* (VU)
- *Bathmocercus cerviniventris* (NT)
- *Bycanistes cylindricus* (NT)
- *Ceratogymna elata* (NT)
- *Illadopsis rufescens* (NT)
- *Lamprotornis cupreocauda* (NT)
- *Malaconotus lagdeni* (NT)
- *Melignomon eisentrauti* (DD)

Bats (5)

Rhinolophus hillorum (VU)
Hipposideros fuliginosus (NT)
Scotonycteris zenkeri (NT)
Hypsugo (crassulus) bellieri (n.a.)
Neoromicia aff. *grandidieri* (n.a.)

Large Mammals (17)

Hexaprotodon liberiensis (EN)
Cercopithecus diana (EN)
Pan troglodytes verus (EN)
Piliocolobus badius (EN)
Loxodonta africana cyclotis (VU)
Cephalophus jentinki (VU)
Syncerus caffer (LR/cd)
Cephalophus dorsalis (LR/nt)
Cephalophus maxwelli (LR/nt)
Cephalophus niger (LR/nt)
Cephalophus ogilbyi (LR/nt)
Cephalophus silvicultor (LR/nt)
Cercocebus atys (LR/nt)
Colobus polykomos (LR/nt)
Procolobus verus (LR/nt)
Tragelaphus euryceros (LR/nt)
Panthera pardus (CITES I)

The IUCN Red List categorizes species based on the degree to which they are threatened. Categories, from less threatened to most threatened, include: Data Deficient (DD, not enough is known to make an assessment), Lower Risk (LR) which includes Conservation Dependent (cd), Near Threatened (nt), and Least Concern (lc, listed but not threatened), Vulnerable (VU), Endangered (EN), and Critically Endangered (CR) (IUCN 2006); n.a.: not assessed by IUCN buy likely to be threatened.

CITES Appendices I, II and III list species afforded different levels or types of protection from over-exploitation (see http://www.cites.org/eng/app/index.shtml).

CONSERVATION RECOMMENDATIONS

North Lorma, Gola and Grebo National Forests all contain a wealth of biodiversity and a significant number of species of conservation concern and each qualifies as an Important Bird Area. Large numbers of the recorded species are restricted to the forests of Upper Guinea. Over 40% of the remaining Upper Guinea forest lies within Liberia and it includes several large tracts of contiguous forest making these among the last refuges for large migrating mammals. For these reasons it is recommended to:

- Raise the status of Grebo National Forest to National Park. Grebo National Forest's close proximity to both Taï National Park in Côte d'Ivoire and Sapo National Park offers an opportunity to create a biological corridor between the two parks.
- Raise the status of North Lorma National Forest to National Park. It is suggested that Wonegizi and Wologizi Mts. also be included in such a park and that the area should be contiguous to the Biosphere Reserve of the Massif du Ziama in Guinea.
- Raise the status of Gola National Forest to National Park and create a transboundary biological corridor with the Gola Forest in Sierra Leone.
- Create a mechanism within which all potential protected areas, especially Grebo, North Lorma and Gola National Forests, can be given blanket protective coverage to allow time for gradual biological, socio-economic and other relevant studies to occur.
- Involve all stakeholders, especially local communities, at an early stage in an open, transparent manner, when establishing forest management plans for these sites.

Executive Summary

INTRODUCTION

Liberia lies entirely within the Upper Guinea forest region that stretches from Guinea to Togo and is part of the Guinean Forests of West Africa Hotspot, making it one of the 34 biologically richest and most endangered terrestrial ecoregions in the world (see map, Myers et al. 2000, Mittermeier et al. 2004). The remaining forests in this region contain exceptionally diverse ecological communities, distinctive flora and fauna, and a mosaic of forest types providing refuge to a number of endemic species (McCullough 2004).

At the Upper Guinean Forest Priority-Setting Workshop in 1999, Gola National Forest and Grebo National Forest were ranked as being 'exceptionally high' and North Lorma National Forest as 'very high' conservation priority areas (Bakarr et al. 2001). Since more than 40% of forest remaining in this hotspot is located in Liberia (Bakarr et al. 2004), the country is key to protecting what is left of the region's fragmented forests.

Historically, Liberia's 9.6 million hectares were completely forested; however, only 36% remains as intact closed forest (2.4 million hectares) or as open forest (1 million hectares) with evidence of recent logging (Bayol and Chevalier 2004). An additional 24% has been altered by agriculture of which nearly 10% is potentially suitable for sustainable forestry. The estimated annual deforestation rates of 1.6% between 1990-2000 and 1.8% between 2000–2005 are higher than that of the total Upper Guinea forest region (1.4% and 1.6%) (FAO 2005).

A large portion of Liberia's forest lies in two large blocks: the evergreen lowland forest in the southeast and the semi-deciduous montane forests in the northwest. Overall, little is known about the country's flora and fauna since few studies have been conducted here.

Liberia currently has two protected areas: Sapo National Park (created in 1983) located in the lowland rainforests of southeastern Liberia and East Nimba Nature Reserve (created in 2003) located in the highest elevations of northeastern Liberia. Prior to the civil war (1989-2003), Sapo National Park was well managed and destined to be the model for all of Liberia's future parks. However, the long war eroded infrastructure and restricted management. Since the end of the war, a concerted effort has been put forth to re-establish conservation, restore the current protected areas and promote sustainable forest management (Waitkuwait and Suter 2001, 2002; Whiteman 2004). In 2002, Conservation International (CI) signed a Memorandum of Understanding (MoU) with the Government of Liberia which proposed seven conservation areas to form the basis of a Liberian protected area network (Conservation International 2002). This would increase the area under protection from 0.2% to 10.6% (Bayol and Chevalier 2004). As Liberia emerges from 14 years of civil war, there will be tremendous pressure on the natural resources to develop the economy (ITTO 2006). A balance between the needs of Liberians and those of their region's imperiled flora and fauna will need to be achieved.

RAP EXPEDITION OVERVIEW AND OBJECTIVES

Conservation International's Rapid Assessment Program (RAP) was created in 1990 to rapidly provide biological information needed to catalyze conservation action and improve biodiversity protection. From November 19 to December 11, 2005, RAP collaborated with CI's Liberia field office and West Africa Program to carry out a biodiversity survey of three sites in Liberia: 1) North Lorma National Forest in the northwest, 2) Gola National Forest in the northwest, and 3) Grebo National Forest in the southeast. These areas still contain large blocks of contiguous low to medium elevation forest and it was deemed important to survey the biodiversity of these areas in order to make recommendations regarding their protection and management.

The RAP survey was part of the larger Liberia Forest Initiative (LFI) which supports efforts to rehabilitate and reform the forest sector in Liberia and harmonize activities associated with these efforts (Whiteman 2004). In addition to collecting data on the sites to strengthen and expand conservation efforts in Liberia, the RAP team worked to build scientific capacity within Liberia's Forestry Development Authority (FDA), the University of Liberia, and local NGOs. Efforts were also made to increase the general public's awareness of their rich flora and fauna.

A RAP team of 21 international and host-country biologists and forestry managers surveyed plants, dragonflies and damselflies, amphibians and reptiles, birds and mammals. International scientists from Belgium, Germany, the Netherlands, Sierra Leone and Swaziland and Liberians from the FDA, the University of Liberia and the Society for the Conservation of Nature in Liberia (SCNL) participated in the study.

RESULTS BY SITE

Coordinates were taken with a Garmin eTrex Venture GPS, map datum WGS 84. See Table 1 for a summary of the number of species recorded at each site.

North Lorma National Forest (19–24 November 2005)

Site 1: 08° 01' 53.6" N 09° 44' 08.6" W

In 1959 the Government of Liberia created the 71,226 hectare North Lorma National Forest (UNEP-WCMC 2006a). It consists of seasonal moist evergreen and semi-deciduous forest. It is situated in northwestern Lofa County near the border with Guinea and constitutes an important forest corridor between the Wologizi and the Wonegizi Mountains. These two mountain ranges, which include Liberia's highest peak, Mt. Wutewe (1424 m), form the most important montane region in Liberia apart from Mt. Nimba. Annual precipitation at North Lorma is approximately 2500 mm and the annual mean temperature is 24.9°C (Chapter 5). Threats to the area include agriculture and hunting (Sambolah 2005).

The camp at Site 1 was situated next to the Lawa River in an open riverine forest habitat. Many smaller streams were present within a slightly hilly landscape. Further from the river, at a higher elevation, the vegetation quickly changed to lower forest with large scattered trees. This site had the least amount of disturbance with the presence of an old overgrown logging road the only noticeable sign of previous human activity.

Significant findings:

- This was the richest site for plants with 266 plant species collected, of which 39 (15%) are endemic to Upper Guinea. One species, *Gardenia nitida*, is a new country record for Liberia. Many different vegetation types were found in close proximity to each other. Next to the Lawa River, species-rich wet forest quickly changed into dry forest and even into completely herbaceous vegetation uphill, whereas in lower areas it gradually changed into swamp forest. Although a logging road was discovered near the camp, logging did not appear to occur in the area.

- Species numbers of dragonflies and damselflies (Odonata) were relatively low due to the unfavorable season. However, the 58 species found are all representative of the Upper Guinean rainforest fauna. *Tetrathemis polleni* marked a new country record for Liberia.

- The recorded herpetofauna consisted of 18 amphibian species and six reptile species. One amphibian species is categorized as Endangered, one as Vulnerable and four as Near Threatened (IUCN 2006, Table 2). Most of these species were very abundant. One amphibian species and two reptile species are listed under CITES. Three amphibian species were new records for the country.

- At this site, 143 species of birds were recorded. Of these, eight are of global conservation concern with two species listed as Vulnerable, five as Near Threatened and one as Data Deficient (IUCN 2006, Table 2). Seven of the 15 restricted-range species (i.e. landbird species which have a global breeding range of less than 50,000 km^2) that make up the Upper Guinea forests Endemic Bird Area (the area from Sierra Leone and southeast Guinea to southwest Ghana that encompasses the overlapping breeding ranges of restricted-range species, Stattersfield et al. 1998) were found during the study. The reserve holds an important proportion of the Upper Guinea endemics and qualifies as an Important Bird Area (IBA, see Birdlife International 2006 for more on IBAs).

- Seven species of bats were found that are restricted to good forest habitat including *Hipposideros fuliginosus*, categorized by IUCN as Near Threatened. A large cave system sheltering over a thousand *Rousettus aegyptiacus* was also found near this site.

- Of the 21 large mammal species recorded, 11 (52%) appear on the IUCN Red List (Table 2). Primates were seen daily and eight species were observed, including the nests of West African Chimpanzees. This was also the only site where African Buffalo was seen.

Gola National Forest (28 November – 4 December 2005)
Site 2: 07° 27' 09.9" N 010° 41'33.2" W
SLC Village: 07° 26' 56.3" N 010° 39' 05.0" W

Gola National Forest was established in 1960 and covers 202,000 hectares (UNEP-WCMC 2006a). It is a seasonal moist evergreen and semi-deciduous forest and is situated in Gborpolu County between the Gola Strict Nature Reserve in Sierra Leone and Kpelle National Forest in Liberia. Annual precipitation at Gola is approximately 2700 mm and the annual mean temperature is 25.4°C (Chapter 5). Threats to the area include logging, hunting and diamond mining (Sambolah 2005).

Two camps were established. The main camp, Site 2, was located within mainly primary forest. The landscape had steep slopes with small, rocky streams. The vegetation was dense in most areas and some huge lianas were present. After it was found that the terrain and closed canopy was hindering sampling efforts for some taxonomic groups, a second camp was established in a clearing at the SLC village. The clearing was once the site of a Spanish Liberia Company (SLC) sawmill, which is now completely destroyed. Illegal small-scale diamond mining was observed just inside the forest.

Significant findings:

- In total, 200 plant species were identified, of which 53 (27%) are endemic to Upper Guinea, including three species endemic to Liberia (*Cephaelis micheliae, Trichoscypha linderi* and *Sericanthe adamii*). A large liana, *Zanthoxylum psammophilum*, not previously recorded west of eastern Côte d'Ivoire, constitutes a new record for Liberia. A *Rhaphiostylis* species likely to be new to science was also discovered. Three saprophytic plant species without chlorophyll were found next to each other at one location. These are not commonly seen and even more rarely in such close proximity.

- Seventy species of Odonata were collected, of which five are Upper Guinean endemics. Two, *Sapho fumosa* and *Trithemis Africana*, are of conservation concern and have a preliminary assessment of Near Threatened. Four species are new country records for Liberia (*Paragomphus nigroviridis, Phyllogomphus moundi, Palpopleura deceptor* and *Trithemis monardi*).

- Thirty amphibians and nine reptiles were recorded, including one Endangered, two Vulnerable and six Near Threatened amphibian species. Two reptile species are listed under CITES (Table 2). Two of the amphibian species constituted new country records for Liberia. The diversity of amphibians and reptiles was higher at this site than at the other two, but included more non-forest species, possibly because of the clearing of forest for diamond mining.

- In total, 145 bird species were found including six Upper Guinea Forest endemics. The Gola Malimbe *Malimbe ballmanni*, categorized as Endangered, was seen on most days. One Vulnerable and four Near Threatened bird species were also noted (Table 2). This site qualifies as an Important Bird Area (IBA).

- A Vulnerable species of bat, *Rhinolophus hillorum*, with a restricted distribution and known from only a few specimens was found. Two additional species that were recorded, *Hypsugo* (*crassulus*) *bellieri* and *Neoromicia* aff. *grandidieri*, are restricted to West Africa, with the latter being a new record for Liberia, and possibly representing a species new to science. In total, 13 bat species and five terrestrial small mammals were noted, including the rarely reported Western Palm Squirrel *Epixerus ebii*.

- Of the 14 species of large mammals recorded, one is listed as Vulnerable and four as Lower Risk/Near Threatened (Table 2).

Grebo National Forest (7 – 11 December 2005)
Site 3: 05° 24' 10.4" N 007° 43' 56.2" W
Jalipo Village: 05° 22' 10.5" N 007° 46' 14.5" W

Grebo National Forest was created in 1960 and covers 260,326 hectares (UNEP-WCMC 2006a). It is a wet evergreen forest situated in the southeast of the country in River Gee County. It is contiguous with the Forêt Classée du Cavally and very close to Taï National Park, both in Côte d'Ivoire, but forest habitat is broken by a narrow strip of dense human settlement and farming on the Ivorian side next to the latter. Annual precipitation at Grebo is approximately 2500 mm and the annual mean temperature is 25.7°C (Chapter 5). Threats to the area include logging and hunting.

Two camps were again established to increase the sampling efforts for some taxonomic groups. The main camp, Site 3, was located in the forest along an old logging road, with a secondary camp at the forest edge in Jalipo Village. Site 3 was a former logging area left untouched for some twenty years and consisting of mainly open, mature secondary forest with isolated huge trees. Some medium-sized sandy streams with a few rocks occurred, as well as large ponds.

Significant findings:

- In total, 220 plant species were recorded, of which 37 (17%) are endemic to Upper Guinea, including a new country record for Liberia (*Elytraria ivorensis*). Two species, a *Drypetes* and a *Leptoderris* are likely to be new to science. The abundant presence of *Psychotria kwewonii* was interesting as it is a recently discovered species occurring in eastern Liberia and southwestern Côte d'Ivoire.

- Of the 63 Odonates collected two are Upper Guinean endemics. *Nesciothemis minor* is a new country record for Liberia.

- The herpetofauna diversity at this site was high, with 30 species of amphibians and six species of reptiles being identified. These consisted mainly of true forest species, with two Vulnerable, ten Near Threatened and possibly one Endangered species (Table 2). Two reptile species are listed under CITES.

- Ten of the 156 bird species recorded are of global conservation concern, with five Vulnerable and five Near Threatened species (Table 2). Nine bird species are Upper Guinea endemics. Additionally, a number of rare and poorly known species were observed including Spot-breasted Ibis *Bostrychia rara*, Congo Serpent Eagle *Urotriorchis spectabilis* and Blue-headed Bee-eater *Merops muelleri*. The site also qualifies as an Important Bird Area (IBA).

- All of the 12 bat species captured in Grebo prefer forested habitat, including *Scotonycteris zenkeri*, a Near Threatened species. *Neoromicia guineensis* is a new country record for Liberia. Five other small mammal species were noted including the scaly-tailed swirrel, *Anomalurus* cf. *pusillus*, which is the third record for West Africa.

- All but one of the 29 large mammal species recorded on the RAP survey were seen in Grebo. Primates were regularly noted including Olive Colobus (*Procolobus verus*) and West African Chimpanzee (*Pan troglodytes versus*). The Red River Hog (*Potamochoerus porcus*) was the only mammal recorded only from this site. Of the 28 large mammal species recorded 14 (50%) are of conservation concern (Table 2). Tracks of Leopard *Panthera pardus* (CITES Appendix I) were observed and large numbers of primates were seen and heard daily.

RESULTS BY TAXON

Plants

We recorded 548 plant species (Table 1) of which 101 (18%) are endemic to the Upper Guinea forest area (Upper Guinea sensu White 1983). The sites in North Lorma and Gola are considered to be healthy mature forest and showed only limited disturbance by human activity that does not, at present, constitute a threat to the vegetation. Grebo was logged about 20 years ago and is now in the process of regeneration and recovering well. We found three plant species endemic to Liberia, *Cephaelis micheliae, Sericanthe adamii* and *Trichoscypha linderi*, and three plant species that were recorded for the first time in the country, *Elytraria ivorensis, Gardenia nitida* and *Zanthoxylum psammophilum*. We also found three plant species likely to be new to science, *Drypetes* sp., *Leptoderris* sp. and *Rhaphiostylis* sp.

Dragonflies and damselflies

We recorded 93 species of dragonflies and damselflies (Table 1). Seven species were recorded in Liberia for the first time. Numbers of species and individuals seemed low, probably because the survey was at the end of the wet season, rather than towards the start. The results nonetheless indicate a healthy watershed in each forest, with limited pollution and streambed erosion. If forest cover and natural stream morphology are retained, the present dragonfly faunas are expected to persist. The most interesting species assemblage was recorded in Gola, including two species of conservation concern (Table 2).

Amphibians and Reptiles

We recorded at least 40 amphibian and 17 reptile species (Table 1). Fifteen amphibians are on the IUCN Red List: two are classified as Endangered, two as Vulnerable, and 11 as Near Threatened (Table 2). We found five species that had not been recorded in Liberia before. For several species records represent large range extensions. Five of the reptile species recorded and one amphibian species are listed under CITES (Table 2). All three forests have a high conservation value as their herpetofauna mainly consists of forest specialists which are endemic to the Upper Guinea forest block.

Table 1. Number of species documented during the RAP survey in the North Lorma, Gola and Grebo National Forests, Liberia.

	All RAP sites in this survey	North Lorma	Gola	Grebo
Plants	548	266	200	220
Dragonflies and damselflies	93	58	70	63
Amphibians	40	18	30	30
Reptiles	17	6	9	6
Birds	211	143	145	156
Bats	22	7	13	12
Small Mammals	9	7	5	5
Large Mammals	29	21	14	28
Total	**969**	**526**	**486**	**520**

Birds

We recorded 211 bird species: 143 at North Lorma, 145 at Gola, and 156 at Grebo (Table 1). Of these, 14 are of conservation concern (eight in North Lorma, six in Gola and 10 in Grebo), amongst which one is classified as Endangered (Gola Malimbe *Malimbus ballmanni*), six as Vulnerable, six as Near Threatened and one as Data Deficient (Table 2). Twelve of the 15 species restricted to the Upper Guinea forests Endemic Bird Area and 136 (or 74%) of the 184 Guinea-Congo forests biome species recorded in Liberia were found during the study. Range extensions or new localities were noted for several species. All three sites qualify as Important Bird Areas (IBA).

Bats and terrestrial small mammals

A total of 182 bats of 22 species were captured (Table 1), representing 37% of the bat species known to occur in Liberia. Species richness was highest at Gola and Grebo, possibly because secondary forest and forest edge was sampled there. North Lorma, where only forest interior was surveyed, had both the lowest capture success and the lowest species richness. Three IUCN Red List species were recorded (Table 2). Bat assemblages in each of the surveyed areas were characterized by forest-dependent species. Not a single species typical of savanna habitats was recorded, indicating high habitat integrity of the National Forests. Three bat species are reported for the first time from Liberia (*Rhinolophus landeri*, *Neoromicia guineensis* and *Neoromicia* aff. *grandidieri*), raising the bat species total for the country to 59. Two species of shrews, one murid rodent, five squirrels and one anomalure (scaly-tailed squirrel) were also recorded, including the rarely reported Western Palm Squirrel *Epixerus ebii* and the Lesser Anomalure *Anomalurus* cf. *pusillus.*

Large mammals

We recorded 29 mammal species including nine primates: 21 in North Lorma, 14 in Gola and 28 in Grebo National Forest (Table 1). Four are listed by the IUCN Red List as Endangered, one as Vulnerable, one as Lower Risk/Conservation Dependant, and nine as Lower Risk/Near Threatened (Table 2). The CITES-listed Leopard *Panthera pardus* was also recorded.

CONSERVATION RECOMMENDATIONS

North Lorma, Gola and Grebo National Forests all contain a wealth of biodiversity and a significant number of species of conservation concern (Table 2). Among the fragmented forests of the Upper Guinea hotspot, Liberia has a great potential for conserving large tracts of contiguous forest that house this flora and fauna. In total, 60 species of conservation concern, as categorized by the IUCN Red List and CITES, were recorded in these forests, a considerably large number. Efforts should be taken to monitor and protect all these species.

Species listed on the IUCN Red List are categorized based on the degree to which they are globally threatened. The data used in the evaluations is objective, based in science and peer reviewed. Categories, from less threatened to most threatened, include: Data Deficient (DD, not enough is known to make an assessment), Least Concern (LC, listed but not threatened), Near Threatened (NT), Vulnerable (VU), Endangered (EN) and Critically Endangered (CR) (IUCN 2006). For species that have not been evaluated since 2001 some categories are slightly different with the category Lower Risk (LR) including Least Concern (lc), Near Threatened (nt) and Conservation Dependent (cd) (IUCN 2006). Species listed by CITES are categorized by how international trade in these species would affect their survival. The Convention provides various levels of trade restrictions based on the Appendix the species is listed under. Appendices from the least restrictive trade to the most restrictive trade include: Appendix III (species not threatened with global extinction), Appendix II (not threatened with extinction but could be if trade is not strictly controlled) and Appendix I (most endangered species, threatened with extinction).

Conservation Priorities

- Raise the status of Grebo National Forest to National Park. Despite human disturbance, 40 animal species of conservation concern were recorded here, the highest number of all three sites visited (Table 2). Many of the bird and amphibian species recorded here have restricted ranges. With nine primate species occurring, primate species diversity at least equals that of nearby Sapo National Park (Waitkuwait 2001). Grebo's close proximity to both Taï National Park in Côte d'Ivoire and Sapo National Park offers an opportunity to create a biological corridor between the two parks. This biological corridor could not only safeguard biodiversity but would also preserve the extensive forest cover that is essential to perpetuate the moist air carried by the southwest monsoon further inland.

- Raise the status of North Lorma National Forest to National Park. In total, RAP scientists found 30 amphibian, reptile, bird and mammal species of conservation concern (Table 2). The variety of habitats and the limited amount of disturbance from human activity make this an area that should be closely monitored so that these habitats remain intact. It is suggested that Wonegizi and Wologizi Mts. also be included within the park and that the area should be contiguous to the Biosphere Reserve of the Massif du Ziama in Guinea. This would constitute one of the most significant protected areas of submontane rainforest in West Africa (for a detailed discussion of the importance of the Massif du Ziama, see Fahr et al. 2006). This mountainous region contains suitable habitat for several cave-roosting bats, many of which have small distribution ranges and are globally threatened.

- Raise the status of Gola National Forest to National Park. At this site RAP scientists recorded 27 animal species of conservation concern (Table 2). Of the three sites surveyed, Gola National Forest contained the highest number of threatened species for bats (three) and odonates (two) and was second highest for amphibians (nine). This site had the highest number (80) of recorded species that are endemic to the Upper Guinea forests. A transboundary biological corridor with the Gola Forest in Sierra Leone could be created allowing migratory animals, such as Forest Elephant, to move between the two countries. Biological surveys should also be conducted at nearby Kpelle National Forest to examine the feasibility of extending the corridor.

- All three sites surveyed qualify as Important Bird Areas (IBAs), which further indicates their importance for biodiversity conservation. IBAs are globally recognized sites for conservation, small enough to be conserved in their entirety and often already part of a protected-area network (BirdLife International 2006). IBAs are designated based on one (or more) of three criteria: 1) Holding significant numbers of one or more globally threatened bird species, 2) Being one of a set of sites that together hold a suite of restricted-range bird species or biome-restricted bird species, and 3) Having exceptionally large numbers of migratory or congregatory bird species (BirdLife International 2006).

General Conservation Recommendations

- Create a mechanism within which all potential protected areas, especially Grebo, North Lorma and Gola National Forests, can be given blanket protective coverage to allow time for gradual biological, socio-economic and other relevant studies to occur. The economic pressures on the natural resources of Liberia are immense and were seen in Gola National Forest where small-scale diamond mining and large scale prospecting are occurring. Many of the species that were recorded during this survey depend on healthy intact forests and if these areas are degraded prior to obtaining protection their survival here could be impacted.

- Involve all stakeholders, especially local communities, at an early stage in an open, transparent manner when establishing forest management plans for these sites.

Table 2. Animal species of conservation concern recorded during the RAP survey (IUCN 2006, UNEP-WCMC 2006).

Taxon	Species Name	Common Name	Conservation Status*	RAP Survey Site		
				North Lorma	Gola	Grebo
Amphibian	*Amnirana occidentalis*	Ivory Coast Frog	EN		x	
Amphibian	*Phrynobatrachus annulatus*		EN	x		
Amphibian	*Phrynobatrachus* cf. *annulatus*		EN	x		
Bird	*Malimbus ballmanni*	Gola Malimbe	EN		x	
Large Mammal	*Hexaprotodon liberiensis*	Pygmy Hippopotamus	EN			x
Primate	*Cercopithecus diana*	Diana Monkey	EN	x		x
Primate	*Pan troglodytes verus*	West African Chimpanzee	EN	x		x
Primate	*Piliocolobus badius*	Western Red Colobus	EN	x		x
Amphibian	*Conraua alleni*	Allen's Slippery Frog	VU		x	x
Amphibian	*Phrynobatrachus villiersi*		VU	x	x	x
Bird	*Agelastes meleagrides*	White-breasted Guineafowl	VU			x
Bird	*Bleda eximius*	Green-tailed Bristlebill	VU			x
Bird	*Criniger olivaceus*	Yellow-bearded Greenbul	VU	x	x	x
Bird	*Lobotos lobatus*	Western Wattled Cuckoo-shrike	VU			x
Bird	*Melaenornis annamarulae*	Nimba Flycatcher	VU			x
Bird	*Picathartes gymnocephalus*	Yellow-headed Picathartes	VU	x		
Bat	*Rhinolophus hillorum*	Upland Horseshoe Bat	VU		x	
Large Mammal	*Loxodonta africana cyclotis*	Forest Elephant	VU	x	x	x
Large Mammal	*Cephalophus jentinki*	Jentink's Duiker	VU			x
Large Mammal	*Syncerus caffer*	African Buffalo	LR/cd	x		
Damselfly	*Sapho fumosa*		NT		x	
Dragonfly	*Trithemis africana*		NT		x	
Amphibian	*Afrixalus nigeriensis*	Nigeria Banana Frog, Banana Tree Frog	NT		x	x

Table 2. *(continued)*

Taxon	Species Name	Common Name	Conservation Status*	RAP Survey Site		
				North Lorma	Gola	Grebo
Amphibian	*Bufo togoensis*		NT	x		x
Amphibian	*Hyperolius chlorosteus*		NT		x	x
Amphibian	*Leptopelis macrotis*	Big-eared Forest Frog	NT			x
Amphibian	*Leptopelis occidentalis*	Taï Forest Tree Frog	NT			x
Amphibian	*Petropedetes natator*		NT		x	
Amphibian	*Phrynobatrachus alleni*		NT	x	x	x
Amphibian	*Phrynobatrachus guineensis*		NT			x
Amphibian	*Phrynobatrachus liberiensis*		NT	x	x	x
Amphibian	*Phrynobatrachus phyllophilus*		NT	x	x	x
Amphibian	*Ptychadena superciliaris*		NT			x
Bird	*Bathmocercus cerviniventris*	Black-headed Rufous Warbler	NT	x		
Bird	*Bycanistes cylindricus*	Brown-cheeked Hornbill	NT	x	x	x
Bird	*Ceratogymna elata*	Yellow-casqued Hornbill	NT	x	x	x
Bird	*Illadopsis rufescens*	Rufous-winged Illadopsis	NT	x	x	x
Bird	*Lamprotornis cupreocauda*	Copper-tailed Glossy Starling	NT	x	x	x
Bird	*Malaconotus lagdeni*	Lagden's Bush-shrike	NT			x
Bat	*Hipposideros fuliginosus*	Sooty Leaf-nosed Bat	NT	x		
Bat	*Scotonycteris zenkeri*	Zenker's Fruit Bat	NT			x
Large Mammal	*Cephalophus dorsalis*	Bay Duiker	LR/nt	x	x	x
Large Mammal	*Cephalophus maxwelli*	Maxwell's Duiker	LR/nt	x	x	x
Large Mammal	*Cephalophus niger*	Black Duiker	LR/nt	x	x	x
Large Mammal	*Cephalophus ogilbyi*	Ogilby's Duiker	LR/nt	x		x
Large Mammal	*Cephalophus silvicultor*	Yellow-backed Duiker	LR/nt			x
Large Mammal	*Tragelaphus euryceros*	Bongo	LR/nt			x
Primate	*Cercocebus atys*	Sooty Mangabey	LR/nt	x	x	x
Primate	*Colobus polykomos*	Western Pied Colobus	LR/nt	x		x
Primate	*Procolobus verus*	Olive Colobus	LR/nt			x
Bird	*Melignomon eisentrauti*	Yellow-footed Honeyguide	DD	x		
Bat	*Hypsugo (crassulus) bellieri*	Bellier's Pipistrelle	n.a.		x	
Bat	*Neoromicia* aff. *grandidieri*	Grandidier's Pipistrelle	n.a.		x	
Amphibian	*Bufo superciliaris*	African Giant Toad, Congo Toad	CITES I	x		
Reptile	*Osteolaemus tetraspis*	African Dwarf Crocodile	CITES I		x	x
Large Mammal	*Panthera pardus*	Leopard	CITES I			x
Reptile	*Kinixys eros*	Forest Hingeback Tortoise	CITES II			x
Reptile	*Kinixys homeana*	Home's Hingeback Tortoise	CITES II	x		
Reptile	*Python sebae*	African Rock Python	CITES II	x		
Reptile	*Varanus ornatus*	Ornate Monitor	CITES II		x	

* The IUCN Red List categorizes species based on the degree to which they are threatened. Categories, from less threatened to most threatened, include: Data Deficient (DD, not enough is known to make an assessment), Lower Risk (LR) which includes Conservation Dependent (cd), Near Threatened (nt), and Least Concern (lc, listed but not threatened), Vulnerable (VU), Endangered (EN), and Critically Endangered (CR) (IUCN 2006); n.a.: not assessed by IUCN but likely to be threatened.

*CITES Appendices I, II and III list species afforded different levels or types of protection from over-exploitation (see http://www.cites.org/eng/app/index.shtml).

- Carry out additional survey work on all taxa at different times of the year for a more comprehensive inventory that would include population estimates and distribution patterns.

- Continue educating local communities that depend on the forest, as well as the general public, on the importance of maintaining healthy, biologically diverse forests and watersheds.

- Enforce existing laws on hunting. Although hunting in national forests is prohibited in Liberia, evidence was found of active poaching in all three forests. Enforcement and education could lead to a diminished bushmeat trade.

- Monitor species of conservation concern. This could be done in collaboration with the government (FDA), NGOs and Liberian universities.

REFERENCES

Bakarr, M., B. Bailey, D. Byler, R. Ham, S. Olivieri and M. Omland (eds.). 2001. From the Forest to the Sea: Biodiversity Connections from Guinea to Togo. Conservation International. Washington, DC.

Bakarr, M., J.F. Oates, J. Fahr, M.P.E. Parren, M.-O. Rödel and R. Demey. 2004. Guinean forests of West Africa, *In:* Mittermeier, R.A., P.R. Gil, M. Hoffman, J. Pilgrim, T. Brooks, C.G. Mittermeier, J. Lamoreux and G.A.B. Da Fonseca (eds.). Hotspots Revisited: Earth's Biologically Richest and Most Endangered Terrestrial Ecoregions. CEMEX / Agrupación Sierra Madre. Mexico City. Pp. 123–130.

Bayol, N. and J.-F. Chevalier. 2004. Current State of the Forest Cover in Liberia: Forest Information Critical to Decision Making. Final report to the World Bank. Forêt Ressources Management. Mauguio, France.

BirdLife International. 2006. Web site: http://www.birdlife.org/action/science/sites.

Conservation International. 2002. Memorandum of Understanding Between the Government of the Republic of Liberia and Conservation International Foundation, signed January 2002, Monrovia, Liberia.

Fahr, J., B.A. Djossa and H. Vierhaus. 2006. Rapid assessment of bats (Chiroptera) in Déré, Diécké and Mt. Béro classified forests, southeastern Guinea; including a review of the distribution of bats in Guinée Forestière. *In:* Wright, H.E., J. McCullough, L.E. Alonso and M.S. Diallo (eds.). A Rapid Biological Assessment of Three Classified Forests in Southeastern Guinea. RAP Bulletin of Biological Assessment 40. Conservation International. Washington, DC. Pp. 168–180, 245–247.

FAO. 2006. Global Forest Resources Assessment 2005. Progress Towards Sustainable Forest Management. FAO Forestry Paper 147. Food and Agriculture Organization of the United Nations. Rome, Italy.

International Tropical Timber Organization (ITTO). 2006. Status of Tropical Forest Management 2005. ITTO Technical Series N° 24. Yokohama, Japan.

IUCN. 2006. 2006 IUCN Red List of Threatened Species. Web site: www.iucnredlist.org.

McCullough, J. (ed.). 2004. A Rapid Biological Assessment of the Forêt Classeé du Pic de Fon, Simandou Range, Southeastern Republic of Guinea. RAP Bulletin of Biological Assessment 35. Conservation International. Washington, D.C.

Mittermeier, R.A., P. Robles Gil, M. Hoffmann, J. Pilgrom, T. Brooks, C.G. Mittermeier, J. Lamoreux and G.A.B. da Fonseca (eds.). 2004. Hotspots Revisited. Earth's Biologically Richest and Most Endangered Terrestrial Ecoregions. CEMEX/Agrupación Sierra Madre, Mexico City.

Myers, N., R.A. Mittermeier, C.G. Mittermeier, G.A.B. da Fonseca and J. Kent. 2000. Biodiversity hotspots for conservation priorities. Nature 403: 853–858.

Sambolah, R. 2005. Report on the rapid faunal surveys of seven Liberian forest areas under investigation for conservation. Fauna and Flora International. Cambridge, UK.

Stattersfield, A.J., M.J. Crosby, A.J. Long and D.C. Wege. 1998. Endemic Bird Areas of the World: Priorities for Biodiversity Conservation. BirdLife International. Cambridge, UK.

UNEP. 2004. Desk Study on the Environment in Liberia. United Nations Environment Programme. Geneva, Switzerland.

UNEP-WCMC. 2006a. World Database on Protected Areas. Web site: http://www.unep-wcmc.org/wdpa.

UNEP-WCMC. 2006b. UNEP-WCMC Species Database: CITES-Listed Species. Web site: http://www.cites.org/eng/resources/species.html.

Waitkuwait, W.E. and J. Suter. (eds). 2001. Report on the establishment of a community-based bio-monitoring programme in and around Sapo National Park, Sinoe County, Liberia. Fauna and Flora International. Cambridge, UK.

Waitkuwait, W.E. and J. Suter. 2002. Report on the first year of operation of a community-based bio-monitoring programme in and around Sapo National Park, Sinoe County, Liberia. Flora and Fauna International. Cambridge, UK.

White, F. 1983. The Vegetation of Africa: A Descriptive Memoir to Accompany the UNESCO/AETFAT/UNSO Vegetation Map of Africa. UNESCO Natural Resources Research 20: 1–356.

Whiteman, A. 2004. The Liberia Forest Initiative. Web site: http://www.fao.org/forestry/site/lfi/.

Chapter 1

Rapid survey of the plants of North Lorma, Gola and Grebo National Forests

Carel C.H. Jongkind

SUMMARY

As the dry season had not really started yet, very few plants were flowering or fruiting during our expedition. The total number of species recorded from the three sites in North Lorma, Gola and Grebo National Forests is 548, however 101 (18%) are endemic to the Upper Guinea forest area (Upper Guinea sensu White). The North Lorma and Gola National Forests are considered to be healthy and mature and show only limited disturbance by human activity, which, at the moment does not cause a clear threat to the vegetation. Grebo National Forest was logged about 20 years ago and is now in the process of regeneration and is recovering well. We found three species endemic to Liberia (*Cephaelis micheliae, Sericanthe adamii* and *Trichoscypha linderi*) and three species recorded for the first time the country (*Elytraria ivorensis, Gardenia nitida* and *Zanthoxylum psammophilum*). Additionally, three species likely to be new to science were found: *Drypetes* sp., *Leptoderris* sp. and *Rhaphiostylis* sp.

INTRODUCTION

Liberia lies almost entirely within the Upper Guinea forest block, which forms the western part of the West African Guinean Forests hotspot, one of the 34 biologically richest and most endangered terrestrial ecoregions in the world (Mittermeier et al. 2004). The Upper Guinea forest as a whole is threatened, and while most other West African countries have lost the majority of their forest cover (e.g. most of the mature forest in neighboring Côte d'Ivoire is already gone), Liberia's forest cover still seems to be quite extensive. Liberia was originally more than 90% forested, and is currently still covered in large part by mature forest. Liberia's forests are, however, increasingly threatened by logging, shifting agriculture, and hunting and mining activities, with logging companies, such as the Oriental Timber Company, recently demonstrating that these forests can disappear in just a few years when large areas are not protected from exploitation.

Most Upper Guinea endemics are concentrated in and around Liberia and species composition varies greatly within the Liberian forest. Important differences exist between the very wet coastal forest of central Liberia and the much drier forest near the border with Guinea. Variation in rain-fall patterns with increasing seasonality from southeast to northwest Liberia also have an important influence on the vegetation. Liberia's botanical richness is thus certainly not adequately protected within Sapo National Park alone and additional protected areas covering a variety of habitats are needed. More biodiversity research is urgently needed to make it possible for the Liberian government to choose the best locations for new protected areas. Most Liberian forests have never been studied by botanists and many undiscovered species are to be expected here.

During the rapid botanical survey of the North Lorma, Gola and Grebo National Forests we did not attempt to compile a complete list of all plant species occurring at the three sites. With approximately 2300 species known from Liberia (Jongkind 2004), including many epiphytes occurring only in the canopy, such a task would have been impossible even if we had spent a month at each site. Because the study sites had rarely been visited by botanists prior to our survey, we were uncertain what to expect. However, as two of the three sites were near to neighboring Côte d'Ivoire and Sierra Leone, where more botanical research has been carried out in the past (e.g. Aké Assi 2001, 2002), we thought it unlikely that we would encounter a large number of unknown species.

METHODS

The North Lorma, Gola and Grebo National Forests were surveyed for six, seven and five days respectively. The three sites were surveyed by walking more or less at random (local guides with detailed knowledge of the forest were unavailable) through the vegetation for most of the daylight hours. We surveyed as many different vegetation types and identifiable vascular plant species as possible. While this was not a particularly scientific based way of working, with the limited time available and the absence of detailed maps, it was the only logical option. Species that were not definitively identified on the spot were collected. These vouchers were dried as soon as possible at the base camp. The drying was done in a special drying press that used hot air produced by a propane cooking stove. These specimens will be kept in the National Herbarium Nederland at the Wageningen University branch, the Netherlands; duplicate fertile specimens will be transferred to other botanical institutes based on the specialists working there. About 80% of the vouchers are now identified to species level.

For western African plant species assessment of the conservation status is very incomplete and represents only a subset of the plant species that are actually threatened. Using data from the IUCN Red List and CITES Appendices would give the wrong indication of the conservation status of the plant species, therefore, no IUCN Red List or CITES status has been listed for this taxonomic group.

RESULTS

As the dry season had not really started, very few plants were flowering or fruiting during our survey. If the dry season had commenced in November as expected, the species lists would have been longer as many plant species are difficult to identify or find when sterile.

Epiphytes are under-represented for all sites. The botanical team had to rely on the relatively few plants that had accidentally fallen from the canopy for collecting specimens of epiphytes. The data from the three study sites are from relatively small areas because the survey work was done on foot from a single base camp at each site. It is not clear whether the vegetation seen was representative of the forests as a whole due to the lack of data from a larger area.

The total number of species recorded from the three sites is 548 (Appendix 1). Of these, 101 (18%) are endemic to the Upper Guinea forest block. Only 21 species (4%) were noted at all three sites, however, it is likely that the real overlap in species composition between the three sites is (much) larger. Many of the species recorded at one or two sites only are known to occur in the general area of the other sites. Although the very different geomorphology of the three sites, combined with differences in climate, in part explains the differences in our species lists, these differences are likely to decrease with increasing research effort.

We found that the three forests have a number of characteristic species in common, namely *Chrysophyllum africanum, Chytranthus carneus, Heinsia crinita, Ruellia primuloides, Copaifera salikounda, Heritiera utilis, Pauridiantha sylvicola, Stelechantha ziamaeana* and *Strephonema pseudocola.* These are all species of wetter forest and the latter five are endemic to Upper Guinea. The overlap of the species lists of North Lorma and Gola National Forests is 49 species, between North Lorma and Grebo National Forests it is 61 species and between Gola and Grebo National Forests it is 49. Too many variables are involved to know if these differences in overlap are statistically solid; it is expected they are not. A larger overlap between North Lorma and Gola National Forests could have been expected, as the range of several species recorded at these localities, such as the common *Schizocolea linderi,* is not known to extend as far east as Grebo National Forest. However, considerable northward and westward range extensions were noted for several species, including *Mendoncia combretoides, Psychotria biaurita, Pyrenacantha klaineana* and *Sericanthe adamii.* Three species, *Elytraria ivorensis, Gardenia nitida* and *Zanthoxylum psammophilum,* were recorded in Liberia for the first time. The presence of the first two is not a surprise because they were already known from close to the Liberian border.

Site 1: North Lorma National Forest

At this site, 265 species were recorded: 231 species were collected and notes on 34 more were taken (Appendix 1). Of these, 38 species (14%) are endemic to the Upper Guinea forest block. The largest families of flowering plants are the Leguminosae, with 26 species, and the Rubiaceae, with 21 species. Twenty-four species of 'ferns and allies' were identified.

Site 1 consisted of a mostly open, species-rich, riverine forest with abundant *Plagiosiphon emarginatus* mixed with many wet evergreen forest species such as *Achyrospermum oblongifolium, Costus deistelii, Cryptosepalum tetraphyllum, Mapania* spp., *Strephonema pseudocola* and *Triphyophyllum peltatum,* the latter mixed with many more widespread forest species. Slightly uphill from the Lawa River, the vegetation quickly changed to lower forest with scattered huge trees which even harboured characteristic dry-forest species like *Gardenia nitida* and *Grewia pubescens.* In some places this vegetation gave way to predominantly herbaceous vegetation with several species of Labiatae and Acanthaceae, such as

Plectranthus epilithicus that are usually found on seasonally wet, rocky areas and occasionally, the succulent *Sansevieria liberica* and the climbing *Asparagus drepanophyllus*. The seasonally dry wind from the north is clearly much stronger here than at the other two sites, and the shallow soil on rocky substrate found at several places also influences the species composition. On such soil in open forest, we found the orchid *Habenaria macrandra*, once with *Oeceoclades maculata* and a *Nervillia* species. In small, rocky streams we often encountered *Anubias gracilis* and several fern species, such as *Bolbitis salicina*. In low areas between the hills several *Raphia palma-pinus* swamps occurred, with other swamp plants like *Halopegia azurea*. It appeared that logging did not occur at this site but there was an old logging road that ended close to our base camp.

Site 2: Gola National Forest

In total 200 species were recorded here: 172 were collected and notes on 28 other species were taken (Appendix 1). Of these, 52 (26%) are endemic to Upper Guinea, including three species known only from Liberia (in bold and underlined in the list). The largest families of flowering plants are the Rubiacaea, with 29 species, and the Leguminosae, with 11 species. Nineteen species of 'ferns and allies' were identified. The species list for this site is not very long because of the difficulty of access compounded by logistical problems during our stay.

The study site was a completely forested area with good evergreen forest species including *Anisophyllea meniaudii, Cola buntingii, Costus deistelii, Delpydora gracilis, Dicellandra barteri, Diospyros chevalieri, Heinsia crinita, Physacanthus batanganus, P. nematosiphon, Renealmia longifolia,* and *Strephonema pseudocola*. The primary forest canopy had an open structure, probably because of the presence of steep slopes. Lower vegetation was dense in most areas and huge lianas were present. Because of the open structure of the forest, many specialised forest undergrowth species, herbs as well as shrubs, occurred. Along the streams the damage to the vegetation caused by rapidly changing water levels was clearly visible. Several species were adapted to this condition; usually shrubs with flexible twigs and narrow leaves like *Rinorea breviracemosa* were collected along fast-flowing parts of the streams. The tree *Stachyothyrsus stapfiana*, which is near-endemic to Liberia, was found several times on sandy banks of the Gba river, in places where the water was flowing more slowly. On rocks in and above the streams the small specialised herb *Argostemma pumilum* was found in abundance. *Anubias gracilis* and ferns like *Bolbitis salicina* were also abundant on such rocks. The streams in this area definitely held the most interesting waterside vegetation among the three study sites.

The deserted village and a few localized diamond-pits did not have an important influence on the vegetation of the area. Most of the area we saw is likely to be too steep and rocky to be suitable for inhabitation or resource exploitation.

A surprise was the presence of *Zanthoxylum psammophilum*, a large liana not previously known from Liberia and not seen west of eastern Côte d'Ivoire before. Also worth mentioning is the presence of three saprophytic species without chlorophyll, *Burmannia congesta, Gymnosiphon longistylum* and *Voyria platypetala*, at one location even found next to each other. These species are not commonly seen and the author had never seen all three together before. The liana *Sericanthe adamii* was recorded for the first time away from Mt. Nimba. This species, as well as *Trichoscypha linderi* and *Cephaelis micheliae*, is at present known only from Liberia. A species of *Rhaphiostylis* found here and at Grebo National Forest is probably new to science.

Site 3: Grebo National Forest

At this site 220 species were recorded: 177 were collected and notes on 43 other species were taken (Appendix 1). Of these, 37 (17%) are endemic to Upper Guinea. The largest families of flowering plants on the list are the Rubiacaea, with 32 species, and the Leguminosae, with 18 species. Sixteen species of 'ferns and allies' were identified.

This area was heavily logged about twenty years ago and has more or less since then been left undisturbed. However, the damage caused by this logging will continue to exercise a strong influence on the forest structure and the species composition for many years to come. Except for one small area, all forest we saw was open, with only isolated huge trees such as *Antiaris toxicaria, Pentaclethra macrophylla, Piptadeniastrum africanum, Sacoglottis gabonensis, Terminalia superba* and *Triplochiton scleroxylon*, which were giving shade to abundant forest re-growth. The forest understory is presumably much poorer in species now than before the logging. Because of the condition of this forest, liana species are well represented in our species list for this site, although large lianas were rare.

The abundant presence of *Psychotria kwewonii* was interesting. It is a recently discovered species occurring in east Liberia and southwest Côte d'Ivoire that is currently being described. A species of *Leptoderris* likely to be new to science was found once at this site as well; it has previously been recorded in the Putu Hills and western Côte d'Ivoire. A *Drypetes* species collected in flower could not be identified and is likely to be new to science, as is a species of *Rhaphiostylis* found also at Gola National Forest.

DISCUSSION

Today most of the mature forest, stretching from coastal Senegal to the border between Ghana and Togo only a century ago, is gone. The rate of disappearance and the percentage that is gone already differs from publication to publication, but no one disagrees with the fact that most is not there anymore. About 35% to 40% of the remaining Upper Guinea forest block is found in Liberia (Poorter et al. 2004). As many species endemic to the western part of the Upper Guinea forest block are found in most or all forests in Liberia and because the Upper Guinea forest as a whole is threatened, all primary and good secondary forest in Liberia is part of a biodiversity hotspot and worth protecting. Hence the challenge is to distinguish and select Liberian forests with a higher or lower conservation priority.

Site 1 is remarkable because of the many different vegetation types occurring in close proximity to each other. Next to the Lawa River, species-rich wet forest quickly changed into dry forest and even into completely herbaceous vegetation uphill, whereas in lower areas it became swamp forest. If this is characteristic for North Lorma National Forest, Site 1 is definitely worth protecting. As most of these vegetation types occur in narrow bands or small isolated areas only, they are probably easily disturbed and can be expected to have difficulty recovering. The vegetation in this area seems to be balancing between two opposites but the presence of many slow-moving species indicate it has been doing so with success for a long time. There are probably not many areas with the same vegetation pattern, in Liberia or elsewhere.

The many steep slopes and the wet climate at Site 2 caused the water levels in the abundant streambeds to change quickly. The vegetation in this area should be protected if only to avoid rapid erosion. We noted important differences in the vegetation from one location to another within the forest, probably due to this special landscape feature, although these were not as important as at Site 1. Gola National Forest has the highest percentage of restricted-range species of the three forests visited, and the site is expected to hold the richest plant diversity. This would make Gola National Forest one of the richer sites within Upper Guinea. None of the three sites is situated in the hyperwet evergreen forest area, where I would expect the highest percentage of Upper Guinea endemic plants, but the site at Gola National Forest is very close to it (Hawthorne and Jongkind 2006).

Several of the species recorded in North Lorma and Gola National Forests are timber species (*Heritiera*, *Entandrophragma* and *Nauclea*) but the steep slopes of these sites would make any resource extraction difficult. Additionally, these trees and the accompanying vegetation provide erosion control.

At Site 3, patches of primary forest need to be examined in order to better evaluate the successive vegetation stages. At present, forest undergrowth and stream-related vegetation appear to be much poorer than in less damaged forest. Based on the limited knowledge we have, the vegetation before the logging probably resembled that of large parts of Taï National Park in Côte d'Ivoire.

CONSERVATION RECOMMENDATIONS

The vegetation in North Lorma and Gola National Forests showed only limited disturbance by human activities which, at present, do not represent a clear threat. However, additional surveys of the forests should be conducted so that a proper baseline can be established. Additionally, these areas should be closely monitored to ensure that forest composition is not altered by human activity.

Years ago, large-scale logging in Grebo National Forest caused damage to the forest, from which it is now recovering well. Our base camp was located on an old road, the re-opening of which would likely reverse such recovery and, if re-opening is not necessary, the road would be better left to revert to forest. In order to conserve the original vegetation, it is important to protect those animals responsible for seed dispersal of the various plant species. Thus, recommendations that are made to protect animal biodiversity are important for the protection of plants as well.

Additional biodiversity research is needed as soon as possible to aid the Liberian government in the selection and designation of new protected areas. Most Liberian forests have never been studied by botanists and, even in places that have been visited before, specialized research can reveal unexpected new species. For instance, an important component of Liberian forest biodiversity is found high up in the trees. Thus, canopy research will surely discover many new epiphytic Orchids and other plants.

REFERENCES

Aké Assi, L. 2001. Flore de la Côte-d'Ivoire: catalogue systématique, biogéographie et écologie. I. Boissiera 57: 1–396. Conservatoire & Jardin botaniques de la Ville de Genève.

Aké Assi, L. 2002. Flore de la Côte-d'Ivoire: catalogue systématique, biogéographie et écologie. II. Boissiera 58: 1–401. Conservatoire & Jardin botaniques de la Ville de Genève.

Hawthorne, W.D. and C.C.H. Jongkind. 2006. Woody Plants of Western African Forests. A guide to the forest trees, shrubs and lianes from Senegal to Ghana. Royal Botanic Gardens, Kew.

Jongkind, C.C.H. and J. Suter. 2004. List of Liberian vascular plants. Advances in Botanical Knowledge of Liberia Supported by the Liberia Forest Re-assessment Project: 7-9 & appendix 1. Fauna and Flora International, Cambridge, UK.

Keay, R.W.J. and F.N. Hepper. 1954–1972. Flora of West Tropical Africa. 2nd edn., Part 1-3. London, Crown Agents for Overseas Governments and Administrations.

Mittermeier, R.A., P. Robles Gil, M. Hoffmann, J. Pilgrom, T. Brooks, C.G. Mittermeier, J. Lamoreux and G.A.B. da Fonseca (eds.). 2004. Hotspots Revisited. Earth's Biologically Richest and Most Endangered Terrestrial Ecoregions. CEMEX/Agrupación Sierra Madre, Mexico City.

Poorter, L., F. Bonger and R.H.M.J. Lemmens. 2004. West African forests: introduction. *In:* L. Poorter et al. (ed.) Biodiversity of West African Forests. Cabi Publishing, Wallingford, UK.

White, F. 1983. The Vegetation of Africa: A Descriptive Memoir to Accompany the UNESCO/AETFAT/UNSO Vegetation Map of Africa. UNESCO Natural Resources Research 20: 1–356.

Chapter 2

Rapid survey of dragonflies and damselflies (Odonata) of North Lorma, Gola and Grebo National Forests, Liberia

Klaas-Douwe B. Dijkstra

SUMMARY

During a rapid survey of the North Lorma, Gola and Grebo National Forests, 93 species of dragonflies and damselflies were found. Seven species were recorded in Liberia for the first time. Numbers of species and individuals seemed low, probably because the survey was at the end of the wet season, rather than towards the start.The results nonetheless indicate a healthy watershed in each forest, with limited pollution and streambed erosion. If forest cover and natural stream morphology are retained, the present dragonfly faunas are expected to persist. The most interesting species assemblage was recorded in Gola National Forest, including two species of conservation concern. Gola National Forest is a major diamond mining area, and the possible beneficial and detrimental impacts of these activities are discussed. Harboring typical examples of a rich Upper Guinea fauna, each forest, and especially Gola National Forest, deserves to be conserved.

INTRODUCTION

Odonata (dragonflies and damselflies) are receiving increasing attention from scientists and the general public. These graceful, colorful creatures are the quintessence of freshwater health. Due to their attractive appearance, dragonflies and damselflies can function as guardians of the watershed. They can be flagships for conservation, not only of water-rich habitats such as wetlands and rainforests, but also of habitats where water is scarce and, therefore, especially vital to the survival of life. Their sensitivity to structural habitat quality (e.g. forest cover, water limpidity) and amphibious habits make Odonata well suited for evaluating environmental change in the long term (biogeography, climatology) and in the short term (conservation biology), both above and below the water surface (Corbet 1999).

Odonata larvae are excellent indicators of the structure and quality of aquatic habitats (e.g. water, vegetation, substrate), while adult Odonata are highly sensitive to the structure of their terrestrial habitats (e.g. degree of shading). As a consequence, Odonata respond strongly to habitat changes, such as those related to deforestation and erosion. Ubiquitous species prevail in disturbed or temporary waters, while habitats like pristine streams and swamp forests harbor a wealth of more vulnerable and local species. Different ecological requirements are linked to different dispersal capacities. Species with narrow niches disperse poorly, while pioneers of temporal habitats (often created by disturbance) are excellent colonizers. For this reason, Odonata have a potential use in the evaluation of habitat connectivity (Clausnitzer 2003, Dijkstra and Lempert 2003).

Odonata possess characteristics distinct from those of relatively well-studied taxonomic groups like plants, birds, mammals and butterflies. Therefore, their study supplements knowledge obtained from these better-known groups. There are also practical advantages to Odonata as environmental monitors. Aquatic habitats, the focal point of their life histories, are easy to locate, and their diurnal activity and high densities make Odonata easy to study. The number of dragonfly species occurring in Africa is manageable, their taxonomy is fairly well resolved, and identification relatively straightforward. Considering the ever-changing nature of the African environment, be it under human, geological or climatic influence, the study of African Odonata constitutes an exciting challenge, as knowledge of their distribution, ecology and phylogeny helps us understand the past and future of a rapidly changing continent.

This was only the second African RAP survey that included Odonata. The first, at Lokutu in Democratic Republic of Congo (Butynski and McCullough *in press*), proved that it is possible to obtain a fair picture of the local diversity within a short period of time. This picture showed a rich and apparently largely natural Odonata fauna, which probably represents high overall aquatic biodiversity. This result contrasted sharply with the impoverished and imperiled fauna and flora found for the other taxonomic groups studied on that RAP survey. Because of their 'information-rich' potential, Odonata might be placed more at the forefront of RAP surveys and conservation policy. Particularly in forest and freshwater ecosystems, an emphasis on odonate research seems beneficial as a baseline for biodiversity and watershed conservation. Sampling these charismatic insects can demonstrate whether present and future conservation actions are protecting freshwater biodiversity. Moreover, the interpretation of survey results has recently been facilitated by the inclusion of Odonata in IUCN's assessment of freshwater biodiversity in western Africa, which summarizes the distribution, habitat, threats and taxonomy of all species (Dijkstra, unpubl.).

The Odonata of the Upper Guinea forest have been fairly well studied. Landmark papers appeared on Sierra Leone (Carfì and D'Andrea 1994), Ghana (O'Neill and Paulson 2001), the Guinean side of Mt Nimba (Legrand 2003) and Taï Forest in Côte d'Ivoire (Legrand and Couturier 1985). The fauna of Liberia is principally known due to Lempert (1988), who surveyed the country (mostly the eastern half) during a total of six months. His thesis is still the most in-depth study of any tropical dragonfly community and includes countless unique observations of reproductive behavior. Lempert recorded between 140 and 150 species, including numerous unnamed species, especially in the Gomphidae. A number of these have probably been described since by Legrand (1992, 2003) and require re-examination. Judging from data from neighboring countries, the true number of species occurring in Liberia should be approximately 200 (Dijkstra and Clausnitzer 2006); about one-fifth of these do not occur east of Nigeria. Lempert's data were analyzed in combination with this author's data from Ghana (Dijkstra and Lempert 2003). This analysis describes the composition of odonate assemblages in running waters in the Upper Guinea rainforest. As running forest waters harbor the larger part of the region's odonate diversity, particularly of range-restricted species, this baseline is an important tool in the interpretation of the data from the present survey.

Despite Lempert's (1988) efforts, large parts of Liberia remain unexplored, in particular the center (e.g. Grand Bassa and River Cess Counties), the southeast (River Gee, Grand Kru, Maryland) and the northwest (Gbarpolu, Lofa). Central Liberia is probably of lesser interest because it is enclosed in Lempert's survey area and relatively deforested. The southeast is interesting because rainfall is spread most evenly over the year and the region is probably nearest to the center of the Upper Guinea rainforest refugium. The northwest has the most diverse terrain, with marked relief and the strongest savannah influences in a country consisting largely of rainforest. It is also the region with the most marked seasons, with distinct wet (May–Oct) and dry (Nov–Apr) seasons. The three national forests (North Lorma, Gola and Grebo) covered by the present survey lie in three previously unstudied counties (Lofa, Gbarpolu and River Gee respectively).

METHODS

North Lorma National Forest was surveyed from 19 to 25 November, Gola National Forest from 27 November to 3 December, and Grebo National Forest from 5 to 11 December 2005. Adult and larval Odonata were observed and caught with a handnet during daylight at freshwater habitats, and details of their ecology and behavior were noted. Identifications were made using Clausnitzer and Dijkstra (in prep.) and additional literature; taxonomy follows Dijkstra and Clausnitzer (in prep.). Relevant name changes from that checklist and other unpublished revisions by the author are provided in the footnotes. Collected specimens will be deposited in the collection of the National Museum of Natural History (Leiden, The Netherlands).

RESULTS

A total of 93 species of Odonata were found, representing 59% of the estimated 158 species known from the country (Appendix 2). Of these, 60% are forest species found only within the Guineo-Congolian realm, with the remaining 40% being widespread non-forest species. Only 31% of the forest species are of more restricted occurrence (i.e. not occurring throughout the realm). Seven species were recorded for the first time in Liberia: *Paragomphus nigroviridis, Phyllogomphus moundi, Nesciothemis minor, Palpopleura deceptor, Tetrathemis polleni, Tramea limbata* and *Trithemis monardi.*

DISCUSSION

Because no research of Odonata had been undertaken prior to this study in the regions visited, any result from these areas greatly supplements the knowledge of the Upper Guinea fauna in general and the Liberian fauna in particular.

Although the total of 93 species seems high, it compares poorly with the result of the RAP in D.R. Congo, where 86 species were found at a single site during half the number of field days. Moreover, in D.R. Congo 72% were forest species found only within the Guineo-Congolian realm (versus 60% in Liberia), with 53% of these being of more restricted occurrence (versus 31% in Liberia) Of the seven species that were recorded for the first time in Liberia, *Paragomphus nigroviridis* is a widespread forest species, while the remaining additions are widespread non-forest species.

The absence of certain expected species, as well as the generally low individual numbers observed, may be explained by seasonality. High and fluctuating water levels are a possible reason why activity of adult dragonflies is low during the transition from wet to dry season. Conditions are then not only challenging for adult dragonflies (e.g. submerged or variably available oviposition substrates; dangerous conditions for emergence), but also for the researcher, whose access to research sites is limited by high water. Moreover, many species may still be in the larval stage at the close of the wet season, because heightened reproductive activity can be expected at the start of the rains when habitat availability increases. Insect numbers generally seemed low during the RAP survey, especially where concentrations would be expected. For instance, very few nocturnal insects were drawn to light, and fruit on the forest floor attracted low numbers of frugivorous butterflies. Insect captures with Malaise traps were also low. The period from February to May is probably the best for recording Odonata.

Of the species found, 17 are rainforest species that do not enter the Congo Basin (mostly ranging east to Nigeria, Cameroon or Gabon); six of these are Upper Guinea endemics (not occurring east of Togo). Of these, *Prodasineura villiersi*, *Phyllomacromia sophia*, *Eleuthemis* sp. n. and *Zygonyx chrysobaphes* are widespread in the Upper Guinea realm. The first was found at all three sites, the second and third in Gola National Forest only, and the fourth in Grebo National Forest. Unlike the Odonata of northern, eastern and southern Africa, those of central and western Africa were not assessed for the global Red List of 2006, as data were fragmented and relatively limited (Dijkstra and Vick 2004). However, the author has recently collated these data and made a regional and preliminary global assessment (Appendix 2). Six Liberian species have globally been assigned the category Near Threatened or higher: *Sapho fumosa* (Near Threatened = NT), *Mesocnemis tisi* (Endangered = EN), *Agriocnemis angustirami* (Vulnerable = VU), *Phyllomacromia funicularioides* (NT), *Neodythemis campioni* (NT) and *Trithemis africana* (NT). Three additional species occur in adjacent Sierra Leone: *Elattoneura dorsalis* (VU), *Pseudagrion mascagnii* (Critically Endangered = CR) and *Orthetrum sagitta* (NT). Of these nine, only two in the lowest category were found during the survey, both in Gola National Forest:

1. *T. africana* is only known from deeply shaded rainforest streams in Sierra Leone, Liberia and Côte d'Ivoire; Liberia must be the species' stronghold.

2. *S. fumosa* is known from a few sites in Senegal and Guinea-Bissau near the border with Guinea, through Sierra Leone to Mt Nimba, where the only previous Liberian record was obtained by Lempert (1988). The species is closely related to *S. ciliata*, *S. bicolor* and *Umma cincta*. All four species were found on the same stream system at Gola National Forest, although they are ecologically segregated. *S. ciliata*, *S. bicolor* and *U. cincta* favor sandy streams, occurring on the sunniest, shadiest and intermediate sections respectively. *S. fumosa* was found only where streams were rather shaded and dominated by rocks, a preference that explains why the species is confined to the more hilly parts of Upper Guinea.

Three additional species found at Gola National Forest are more widespread in western Africa, ranging east to Cameroon, but have been recorded only locally: *Phyllogomphus moundi*, *Tetrathemis godiardi* and *Trithemis basitincta*. These results indicate that from an odonatological perspective Gola National Forest was the most interesting site. Of the 17 western African species mentioned above, only seven were found at North Lorma National Forest, compared to 14 in Gola National Forest and 12 in Grebo National Forest.

Although deforestation and subsequent alteration of waterbodies (e.g. erosion, siltation) seem to be the only potential threats to Odonata in North Lorma and Grebo National Forests, diamond mining may be detrimental also in Gola National Forest. Small-scale activities that do not open up the canopy appear beneficial. Stagnant waterbodies are comparatively scarce in rainforest, and partly overgrown pits filled with leaf-litter create new habitat. *Tetrathemis godiardi* is the most obvious beneficiary; both territorial and emerging individuals were found at abandoned pits under closed canopy. Open pits are colonised by many species that would otherwise find no or almost no habitat in the area, but these are all well-dispersing species that dominate savannah faunas throughout Africa. The drainage of the mines leads to increased turbidity, and probably siltation of streams, the former reducing visibility for larvae, the latter changing the substrate. Reduced motion and increased insolation of water in open pits also affects the flow, oxygen and temperature regimes of drainage streams. One such stream in Gola National Forest, which was rocky and therefore suitable for *S. fumosa*, held very low numbers of that species in comparison to a pristine stream, but observations are too limited to draw conclusions.

CONSERVATION RECOMMENDATIONS

Odonates were the only invertebrate group included in the RAP. Unlike some other taxonomic groups studied, they are not actively exploited by man and are strongly tied to water. They therefore serve to assess the more indirect anthropogenic disturbance—the gradual alteration of the environment. As expected, all three studied forests harbor odonate assemblages that are representative of the Upper Guinea rainforest fauna. The forest stream assemblages found match those described by Dijkstra and Lempert (2003), suggesting healthy watersheds, with limited degrees of pollution and streambed erosion. As long as forest cover and natural stream morphology are retained, the existing dragonfly fauna is expected to persist. Considering the threats to the Upper Guinea rainforest, it is recommended that the three forests and the watersheds they protect be conserved. This recommendation especially concerns Gola National Forest, which had the most interesting dragonfly fauna, including two species of conservation concern (*Sapho fumosa, Trithemis africana*). The additional threat of diamond mining may jeopardize the aquatic biodiversity in Gola National Forest. Minimizing the outflow of mining water into the stream systems may reduce the possible negative effect of those activities.

REFERENCES

Aguesse, P. 1968. Quelques Odonates récoltés en Sierra Leone. Bull. Inst. fond. Afr. noire, sér. A, 30: 518–534.

Butynski, T. M. and J. M. McCullough (editors). *In press.* A rapid biological assessment of Lokutu, Democratic Republic of Congo. RAP Bulletin of Biological Assessment 46. Conservation International, Arlington, VA, USA.

Carfì, S. and M. D'Andrea. 1994. Contribution to the knowledge of odonatological fauna in Sierra Leone, West Africa. Problemi Attuali di Scienza e di Cultura 267: 111–191.

Clausnitzer, V. 2003. Dragonfly communities in coastal habitats of Kenya: indication of biotope quality and the need of conservation methods. Biodiversity and Conservation 12: 333–356.

Clausnitzer, V. and K.-D.B. Dijkstra. In prep. The dragonflies of Eastern Africa (Odonata), an identification key. Studies in Afrotropical Zoology.

Corbet, P.S. 1999. Dragonflies: Behaviour and Ecology of Odonata. Harley Books, Colchester.

Dijkstra, K.-D.B. and V. Clausnitzer. 2006. Thoughts from Africa: how can forest influence species composition, diversity and speciation in tropical Odonata? *In*: Cordero Rivera, A. (ed.). Forests and Dragonflies. Pensoft Publishers, Sofia.

Dijkstra, K.-D.B. and V. Clausnitzer. In prep. An annotated checklist of the dragonflies (Odonata) of Eastern Africa, with critical lists for Ethiopia, Kenya, Malawi, Tanzania and Uganda, new records and taxonomic notes.

Dijkstra, K.-D.B. and J. Lempert. 2003. Odonate assemblages of running waters in the Upper Guinean forest. Archiv für Hydrobiologie 157: 397–412.

Dijkstra, K.-D.B. and G.S. Vick. 2004. Critical species of Odonata in western Africa. *In*: Clausnitzer, V. & R. Jödicke (eds.). Guardians of the Watershed. Global status of dragonflies: critical species, threat and conservation. Int. J. Odonatol. 7: 229–238.

Legrand, J. 1992. Nouveaux Gomphidae afrotropicaux. Descriptions préliminaires. (Odonata, Anisoptera). Rev. fr. Entomol. (N.S.) 14 (4): 187–190.

Legrand, J. 2003. Les Odonates du Nimba et de sa région. *In*: M. Lamotte and R. Roy: Le peuplement animal du mont Nimba (Guinée, Côte d'Ivoire, Liberia). Mém. Mus. natl. Hist. nat. 190: 231–310.

Legrand, J. and G. Couturier. 1985. Les Odonates de la forêt de Taï (Côte d'Ivoire). Premières approches: faunistique, répartition écologique et association d'espèces. Rev. Hydrobiol. trop. 18 (2): 133–158.

Lempert, J. 1988. Untersuchungen zur Fauna, Ökologie und zum Fortpflanzungsverhalten von Libellen (Odonata) an Gewässern des tropischen Regenwaldes in Liberia, Westafrika. Diplomarbeit, Friedrich-Wilhelms Universität, Bonn.

Marconi, A. and F. Terzani. s.d. Odonati della Sierra Leone. Unpublished.

O'Neill, G. and D.R. Paulson. 2001. An annotated list of Odonata collected in Ghana in 1997, a checklist of Ghana Odonata, and comments on West African odonate biodiversity and biogeography. Odonatologica 30: 67–86.

Pinhey, E. 1984. A checklist of the Odonata of Zimbabwe and Zambia. Smithersia 3: 1–64.

Chapter 3

Rapid survey of amphibians and reptiles of North Lorma, Gola and Grebo National Forests

Annika Hillers and Mark-Oliver Rödel

SUMMARY

During a herpetological survey of three national forests in northwestern (North Lorma, Gola) and southeastern (Grebo) Liberia we recorded at least 40 amphibian and 17 reptile species. Fifteen amphibians are on the IUCN Red List: 11 are classified as Near Threatened, two as Vulnerable and two as Endangered. We observed five species that had not been recorded in Liberia before. For several species our findings represent large range extensions. Five of the reptile species recorded and one amphibian are listed under CITES. All three forests have a high conservation value as their herpetofauna mainly consists of forest specialists which are endemic to the Upper Guinea forest block.

INTRODUCTION

Liberia is assumed to harbour a high biodiversity and to be one of the richest countries in animal and plant species in West Africa (Bakarr et al. 2001). The diversity of amphibians and reptiles is also thought to be extremely high and to comprise a great number of Upper Guinea endemics. The eastern part of the country (Cape Palmas) in particular is assumed to be very rich in species since it was a rainforest refugium in northern glacial times, which were dry periods in Africa (e.g. Sosef 1994). However, while the herpetofauna of neighboring Guinea and Côte d'Ivoire is relatively well documented (e.g. Guibé and Lamotte 1958a, 1958b, 1963; Schiøtz 1967, 1968; Böhme 1994a, 1994b; Rödel and Bangoura 2004; Rödel et al. 2004; Greenbaum and Carr 2005) and surveys were conducted in Sierra Leone's forests in 2005 (Hillers et al. in prep.), there are almost no recent herpetological data available for Liberia.

Whereas most other West African countries have lost most of their forest cover (e.g. more than 80% of Côte d'Ivoire's forests have been logged during the last 30 years: Chatelain et al. 1996), Liberia's forests seem to still be quite extensive. They are, however, increasingly threatened by logging, shifting agriculture, hunting and mining activities. Therefore data on species' occurrence in Liberia and, more specifically, their distributions within the country, are urgently needed.

Amphibians and reptiles are not only important with regard to biodiversity. They are extremely sensitive to habitat changes, which qualifies them as excellent bio-indicators. The composition of amphibian assemblages may indeed be indicative of the quality of a habitat (compare Rödel and Branch 2002; Rödel and Ernst 2003; Ernst and Rödel 2005; Ernst et al. 2006). Based on this knowledge, conservation recommendations can be made.

METHODS

The RAP survey was carried out between 16 November and 14 December 2005, which is the end of the rainy season and the beginning of the dry season, and hence not the most appropriate period to search for amphibians and reptiles.

The first study area, North Lorma National Forest, situated in northwestern Liberia (surveyed from 19–24 November 2005) was characterized by primary forest crossed by a large river and many smaller streams on slightly hilly terrain. There was no obvious anthropogenic disturbance to the investigated locations and habitat variation was minimal.

The second study site, Gola National Forest (28 November–4 December 2005) was mainly characterized by primary forest with rocky streams on very hilly terrain with several indications of anthropogenic impact, including old mining areas and a miners' camp.

The third site, Grebo National Forest (7–11 December 2005) was characterized by a 20-year old, mature secondary forest, but sites investigated contained some remaining primary forest. The actual study site was situated in a previously logged area and next to an old logging road. Aquatic sites included sandy streams with a few stones and rocks, and large pools.

Coordinates were taken with a hand-held GPS receiver (Garmin 12XL; WGS 84) and are listed in Appendix 3.

During our survey we concentrated on amphibians, as there are standardized methods to investigate these, while reptile records were only collected opportunistically. Amphibians were mainly recorded during encounter surveys, conducted both during the day and at night, by up to three people. Searching techniques included visual scanning of the terrain, investigation of potential refuges and acoustical monitoring (see Heyer et al. 1994; Rödel and Bangoura 2004; Rödel and Ernst 2004). Despite the fact that we experienced some rain at the different study sites (with almost daily, heavy rain at the last site), the number of calling males (indicating reproductive activity) was limited. A higher calling activity was observed at pools within the forest, but these did not occur at all study sites.

In addition to visual and acoustic monitoring, drift fences and pitfall traps (15 m of drift fence with five buckets), as well as drift fences and funnel traps (10 m of drift fence with eight funnel traps) were installed at all study sites (four nights per site). Only in Grebo National Forest did the trapping add a single amphibian species to our list. These results are therefore not reported in detail.

As our sampling design provides only qualitative and semi-quantitative data we estimated species richness and sampling efficiency with the Chao 2 and Jackknife 1 estimators (software EstimateS: Colwell 2005). These estimators are incidence based, using the presence/absence data of the daily species lists (15 days) for 40 species. The sampling effort was measured in man-hours spent searching at each study site and it was assumed that this effort was the same for each habitat. To obtain quantitative data, mark-recapture experiments along standardized transects or on well-defined plots would have been necessary. These techniques could not be used due to the limited time available at each site.

The nomenclature of amphibians follows the taxonomy by Frost (2004); changes according to Frost et al. (2006) are listed in Appendix 4. For reptiles, the nomenclature follows Uetz et al. (1995). Some voucher specimens were anesthetized and euthanized in a chlorobutanol solution (amphibians) or ether (reptiles) and preserved in 70% ethanol. Voucher specimens have been deposited at Mark-Oliver Rödel's collection at the University of Würzburg (MOR); some of these will be transferred to natural history museums. Tissue samples (toe tips) were preserved in 95% ethanol. These samples have been stored at the Institute for Biodiversity and Ecosystem Dynamics at the University of Amsterdam, the Netherlands (Appendix 5).

RESULTS

We recorded at least 40 amphibian and 17 reptile species at the different study sites. A list of amphibian species with record sites, known habitat preference, distribution area and IUCN Red List category (IUCN et al. 2004) is given in Appendix 3. The list of reptile species with record sites is given in Appendix 6.

Twelve (30%) of the recorded amphibian species have a known distribution range that exceeds Western Africa. Eight species (20%) are restricted to West Africa, while the majority (19 species, 47.5%) are endemic to the Upper Guinea forest bloc. Further genetic analyses will clarify if one specimen can be referred to a known species (*Phrynobatrachus annulatus*) or if it is new to science and then probably a Liberian endemic (see below). The majority of the amphibians recorded are typical forest specialists, with some of them also being tolerant of farmbush areas (Appendix 4). Only a few prefer savannah and/or farmbush habitats. The occurrence of the latter in a forest is an indication of environmental disturbance (see Rödel and Branch 2002).

In North Lorma National Forest we recorded 18 amphibian and six reptile species, in Gola National Forest 30 amphibian and nine reptile species and in Grebo National Forest 30 amphibian and six reptile species.

North Lorma National Forest's amphibian community was dominated by true primary forest species, including four Near Threatened, one Vulnerable and one Endangered species. Most of these species were all very abundant within the study area. The presence of *Bufo maculatus* was a sign of some habitat disturbance.

Due to the presence of old mining areas, we observed a higher number of invasive species, i.e. species that normally do not occur within forest, in Gola National Forest (e.g. *Hoplobatrachus occippitalis, Afrixalus dorsalis*). The rocky streams represented a special habitat for species that were not or only rarely recorded in the other two forests.

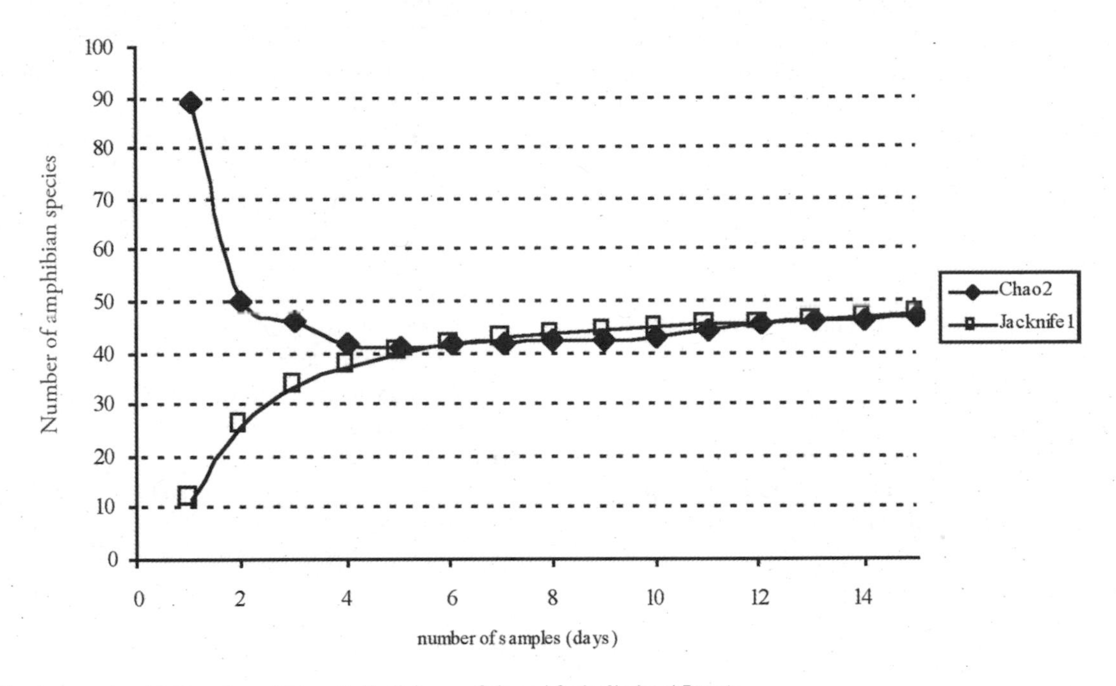

Figure 3.1: Estimated species richness of amphibians in North Lorma, Gola and Grebo National Forests.

We found six Near Threatened, two Vulnerable and one Endangered species in Gola National Forest.

A similar situation occurred in Grebo National Forest, where streams and big pools harboured species characteristic of these habitats. These species were not found at North Lorma National Forest, where similar habitats did not occur. In Grebo National Forest we recorded ten Near Threatened and two Vulnerable species. Surprisingly, these were mainly true forest species, although the study area mostly consisted of mature secondary forest. Genetic analysis will show if the Grebo National Forest community additionally includes one Endangered species.

The two incidence-based species richness estimators calculated that additional amphibian species are likely to occur within the three study sites. Both the Jacknife 1 and the Chao 2 estimator computed a species richness of 47 amphibian species (Figure 3.1). The recorded number of species thus corresponds to 85% of the estimated species richness.

Several amphibian species were recorded for the first time in Liberia: *Bufo superciliaris, Astylosternus occidentalis, Phrynobatrachus villiersi* (Vulnerable), *Chiromantis rufescens* and *Afrixalus nigeriensis* (Near Threatened). Range extensions were noted for *Phrynobatrachus plicatus, Ptychadena aequiplicata* and the Endangered *Amnirana occidentalis.*

DISCUSSION

Although it rained heavily on a few days during the survey period, there were few calling males within the forest. Probably the reproductive period of most species had already come to an end. It is likely that some species were therefore overlooked. This may have led to an underestimation of species richness. Another factor contributing to this underestimation is that we covered only a small area within the three forests and did not visit all the existing habitat types (e.g. inselbergs or Grebo National Forest's primary forest). These habitats are likely to harbour additional frog species. This is also illustrated by the fact that during 15 days of observation we continuously detected additional species.

We recorded all Upper Guinea forest amphibians dependant on lotic forest habitats (*Astylosternus occidentalis, Bufo togoensis, Cardioglossa leucomystax, Conraua alleni, Hyperolius chlorosteus, Petropedetes natator*) as well as some species typical of lentic forest habitats (e.g. *Chiromantis rufescens, Phlyctimantis boulengeri, Phrynobatrachus plicatus*). Five reptile species (African Dwarf Crocodile *Osteolaemus tetraspis,* Rock Python *Python sebae,* Monitor Lizard *Varanus ornatus,* and two tortoises *Kinixys erosa* and *K. homeana*) as well as the toad *Bufo superciliaris* are threatened and protected by international law.

CONSERVATION RECOMMENDATIONS

In all three national forests the observed amphibian community is of high conservation value, comprising typical forest assemblages with only a few invasive farmbush species. Many of the recorded species are Near Threatened and some are Endangered or Vulnerable.

Some Near Threatened species (*Phrynobatrachus alleni, Phrynobatrachus liberiensis, Phrynobatrachus phyllophilus*) were extremely abundant in North Lorma and this habitat thus seemed quite healthy. This amphibian community was dominated by true primary forest species, but the presence of e.g. *Bufo maculatus* was a signal of disturbance. The Vulnerable (*Phrynobatrachus villiersi*), Endangered (*Phrynobatrachus annulatus*) and Near Threatened species mentioned live in extremely different habitat types. The variety of habitats in North Lorma National Forest should therefore be conserved and prevented from any further alteration.

In Gola National Forest, former mining activities created disturbed habitats, which led to the presence of invasive savannah species. However, some of these new habitats were also favourable to forest species (e.g. *Afrixalus nigeriensis, Phrynobatrachus fraterculus, Chiromantis rufescens*), as, for example, old mining ponds were used as breeding sites by various frog species. Possible contamination from former mining that could affect amphibians, such as quicksilver residues, could not be examined during our survey. The many rocky streams in Gola National Forest represent a typical habitat for a number of frogs, including Near Threatened (*Hyperolius chlorosteus, Petropedetes natator*), Vulnerable (*Conraua alleni*) and Endangered (*Amnirana occidentalis*) species. Therefore, this forest, and especially its aquatic habitats, should be protected and preserved from further disturbance.

In Grebo National Forest, the anthropogenic alteration is very obvious. However, a high number of typical forest species occurred in the mature secondary forest, including ten Near Threatened (e.g *Bufo togoensis, Afrixalus nigeriensis, Leptopelis occidentalis, Phrynobatrachus guineensis*) and two Vulnerable species (*Conraua alleni, Phrynobatrachus villiersi*). Further genetic analysis will show if one frog can be correctly assigned to the Endangered *Phrynobatrachus annulatus*. Additionally, we assume Grebo National Forest to be very important for conservation because of its proximity to Taï National Park in Côte d'Ivoire. The latter is one of the most species-rich West African sites for amphibians (Rödel 2000) and harbours highly specialized and endemic forest species (Perret 1988; Rödel and Ernst 2000, 2001; Rödel et al. 2003). Since both sites were connected in the past, several threatened species and species considered endemic to Taï National Park are likely to occur in Grebo National Forest as well. This might be the case for *Kassina lamottei* and *Bufo taiensis*. Grebo National Forest could likely play a major role by connecting Taï and Sapo National Parks as a kind of stepping stone.

Further surveys are highly recommended for all three forests. These should predominantly take place during the rainy season, when amphibians and reptiles are more active. In Grebo National Forest these studies should focus on the primary forest habitat. Intensive research will lead to a better knowledge of the existing herpetofauna and its assemblage composition, population size and distribution patterns. This is especially important for species of conservation concern. Further studies will also lead to a better understanding of the importance of certain habitats, which will help in developing reliable and detailed conservation recommendations.

REFERENCES

Bakarr, M., B. Bailey, D. Byler, R. Ham, S. Olivieri and M. Omland. 2001. From the Forest to the Sea: Biodiversity Connections from Guinea to Togo. Conservation International. Washington DC.

Böhme, W. 1994a. Frösche und Skinke aus dem Regenwaldgebiet Südost-Guineas, Westafrika. I. Einleitung; Pipidae, Arthroleptidae, Bufonidae. Herpetofauna 16 (92): 11–19.

Böhme, W. 1994b. Frösche und Skinke aus dem Regenwaldgebiet Südost-Guineas, Westafrika. II. Ranidae, Hyperoliidae, Scincidae; faunistisch-ökologische Bewertung. Herpetofauna 16 (93): 6–16.

Chatelain, C., L. Gautier and R. Spichiger. 1996. A recent history of forest fragmentation in southwestern Ivory Coast. Biodiv. Conserv. 5: 783–791.

Collwell, R.K. 2005. EstimateS Version 6.0b. Statistical estimation of species richness and shared species from samples. Website: viceroy.eeb.uconn.edu/estimates.

Ernst, R., K.E. Linsenmair and M.-O. Rödel. 2006. Diversity erosion beyond the species level: Dramatic loss of functional diversity after selective logging in two tropical amphibian communities. Biol. Conserv. 133: 143–155.

Ernst, R. and M.-O. Rödel. 2005. Anthropogenically induced changes of predictability in tropical anuran assemblages. Ecology 86: 3111–3118.

Frost, D.R. 2004. Amphibian species of the World: an online reference. Version 3.0. American Museum of Natural History, New York. Website: research.amnh.org/herpetology/amphibia/index.html. (January 6th, 2006).

Frost, D.R., T. Grant, J. Faivovich, R.H. Bain, A. Haas, C.F.B. Haddad, R.O. De Sá, A. Channing, M. Wilkinson, S.C. Donnellan, C.J. Raxworthy, J.A. Campbell, B.L. Blotto, P. Moler, R.C. Drewes, R.A. Nussbaum, J.D. Lynch, D.M. Green and W.C. Wheeler. 2006. The Amphibian tree of life. Bull. Am. Mus. Nat. Hist. 297: 1–370.

Greenbaum, E. and J.L. Carr. 2005. The Herpetofauna of Upper Niger National Park, Guinea, West Africa. Sci. Papers, Nat. Hist. Mus. Univ. Kansas 37: 1–21.

Guibé, J. and M. Lamotte. 1958a. La reserve naturelle intégrale du Mont Nimba. XII. Batraciens (sauf *Arthroleptis, Phrynobatrachus* et *Hyperolius*). Mém. Inst. fond. Afr. noire, sér. A, 53: 241–273.

Guibé, J. and M. Lamotte. 1958b. Morphologie et reproduction par développement direct d'un anoure du Mont Nimba, *Arthroleptis crusculum* Angel. Bull. Mus. natl. Hist. nat., 2e sér, 30: 125–133.

Guibé, J. and M. Lamotte. 1963. La reserve naturelle intégrale du Mont Nimba. XXVIII. Batraciens du genre Phrynobatrachus. Mém. Inst. fond. Afr. noire, sér. A, 66: 601–627.

Heyer, W.R., M.A. Donnelly, R.W. McDiarmid, L.-A.C. Hayek and M.S. Foster. 1994. Measuring and Monitoring Biological Diversity. Standard Methods for Amphibians. Biological Diversity Handbook Series. Smithsonian Institution Press. Washington, DC.

IUCN, Conservation International and Nature Serve. 2004. Global Amphibian Assessment. Website: globalamphibians.org. (January 6th, 2006).

Perret, J.-L. 1988. Les espèces de *Phrynobatrachus* (Anura, Ranidae) à éperon palpépral. Arch. Sci. 41: 275–294.

Rödel, M.-O. 2000. Les communautés d'amphibiens dans le Parc National de Taï, Côte d'Ivoire. Les anoures comme bio-indicateurs de l'état des habitats. *In*: Girardin, O., I. Koné and Y. Tano (eds.). Etat des recherches en cours dans le Parc National de Taï (PNT). Sempervira, Rapport du Centre Suisse de la Recherche Scientifique, Abidjan, 9: 108–113.

Rödel, M.-O. and A.C. Agyei. 2003. Amphibians of the Togo-Volta highlands, eastern Ghana. Salamandra 39: 207–234.

Rödel, M.-O. and M.A. Bangoura. 2004: A conservation assessment of amphibians in the Forêt Classée du Pic de Fon, Simandou Range, southeastern Republic of Guinea, with the description of a new Amnirana species (Amphibia Anura Ranidae). Trop. Zool. 17: 201–232.

Rödel, M.-O. and W.R. Branch. 2002. Herpetological survey of the Haute Dodo and Cavally forests, western Ivory Coast, Part I: Amphibians. Salamandra 38: 245–268.

Rödel, M.-O. and R. Ernst. 2000. *Bufo taiensis* n. sp., eine neue Kröte aus dem Taï Nationalpark, Elfenbeinküste. Herpetofauna 22 (125): 9–16.

Rödel, M.-O. and R. Ernst. 2001. Description of the tadpole of *Kassina lamottei* Schiøtz, 1967. J. Herpetol. 36: 561–571.

Rödel, M.-O. and R. Ernst. 2003. The amphibians of Marahoué and Mont Péko National Parks, Ivory Coast. Herpetozoa 16: 23–39.

Rödel, M.-O. and R. Ernst. 2004. Measuring and monitoring amphibian diversity in tropical forests. I. An evaluation of methods with recommendations for standardization. Ecotropica 10: 1–14.

Rödel, M.-O., J. Kosuch, M. Veith and R. Ernst. 2003. First record of the genus *Acanthixalus* Laurent 1944 from the Upper Guinean rain forest, West Africa, including the description of a new species. J. Herpetol. 37: 43–52.

Rödel, M.-O., M.A. Bangoura and W. Böhme. 2004. The amphibians of south-eastern Republic of Guinea (Amphibia: Gymniphiona, Anura). Herpetozoa 17: 99–118.

Schiøtz, A. 1967. The treefrogs (Racophoridae) of West Africa. Spolia zool. Mus. haun. 25: 1–346.

Schiøtz, A. 1968. On a collection of amphibia from Liberia and Guinea. Vidensk. Medd. dansk naturh. Foren. 131: 105–108.

Schiøtz, A. 1999. Treefrogs of Africa. Edition Chimaira. Frankfurt/M.

Sosef, M.S.M. 1994. Refuge begonias: taxonomy, phylogeny and historical biogeography of *Begonia* sect. *Loasibegonia* and sect. *Scutobegonia* in relation to glacial rain forest refuges in Africa. Studies in Begoniaceae 5. Wageningen Agricultural University Papers.

Uetz, P., R, Chenna, T. Etzold and J. Hallermann. 1995. The EMBL reptile database. Website: www.reptile-database.org (January 6th, 2006).

Chapter 4

Rapid survey of the birds of North Lorma, Gola and Grebo National Forests

Ron Demey

SUMMARY

During 20 days of field work in three Liberian National Forests, between 19 November and 11 December 2005, 211 bird species were recorded: 143 at North Lorma, Lofa County, 145 at Gola, Gborpolu County, and 157 at Grebo, River Gee County. Of these, 14 are of conservation concern (eight in North Lorma, six in Gola and 10 in Grebo), amongst which one is classified as Endangered (Gola Malimbe *Malimbus ballmanni*), six as Vulnerable, six as Near Threatened and one as Data Deficient. Twelve of the 15 species restricted to the Upper Guinea forests Endemic Bird Area and 136 (or 74%) of the 184 Guinea-Congo forests biome species recorded in Liberia were found during the study. Range extensions or new localities were noted for several species. All three sites qualify as Important Bird Areas. Considering the high conservation value of these forests, it is recommended that further surveys be conducted in order to complete avifaunal data.

INTRODUCTION

Birds have been proven to be useful indicators of biological diversity of a site, because they occur in most habitats on land throughout the world and are sensitive to environmental change. Their taxonomy and global geographical distribution are relatively well known in comparison to other taxa (ICBP 1992). The conservation status of most species has been reasonably well assessed and is being regularly updated (BirdLife International 2000, 2004). This permits rapid analysis of the results of an ornithological study and the presentation of conservation recommendations. Birds are also among the most charismatic species, which can facilitate the acceptance of the necessity to implement protective measures by policy makers and stakeholders. Hotspots for birds are generally of importance for plants or other animals as well (ICBP 1992).

As West African forests are rapidly disappearing, the survival of the birds of the Upper Guinea forests is becoming increasingly dependent on ever fewer areas. Despite a number of surveys conducted in the region in recent or relatively recent years (e.g. Allport et al. 1989; Gartshore 1989; Gartshore et al. 1995; Demey and Rainey 2004, 2005; Rainey and Asamoah 2005), the avifaunas of the majority of these forests are still inadequately known.

The most recent and extensive study of the Liberian avifauna is that of Gatter (1997), on which the selection of Liberia's nine Important Bird Areas (IBAs) by Robertson (2001) was largely based. However, site-specific avifaunal information is scarce and in many cases the presence of species at sites that were selected as IBAs was inferred from the species' known distributions in the region or in areas adjacent to the site (Robertson 2001). Much thus remains to be learned on the precise distribution of species and the ornithological importance of certain sites. In view of the ongoing forest destruction, updating the scarce existing information is also indispensable.

Liberia lies almost entirely within the Upper Guinea forest block, which forms the western part of the West African Guinean Forests hotspot, one of the 34 biologically richest and most endangered terrestrial ecoregions in the world (Mittermeier et al. 2004). The Upper Guinea forests have also been identified as an Endemic Bird Area with the highest priority ranking for conservation action, based on its biological importance and current threat level (Stattersfield et al. 1998). All three forests visited during the present survey lie wholly or in part in IBAs: North Lorma National Forest lies entirely within the Wologizi mountains IBA (LR001), part of Gola National Forest is included in Lofa-Mano IBA (LR003), which is contiguous with the Gola Forest Reserves IBA on the other side of the border in Sierra Leone, and part of Grebo National Forest forms the Cavalla River IBA (LR009), situated on the frontier with Côte d'Ivoire, in the extreme east of the country and lying close to Côte d'Ivoire's Taï National Park IBA (Robertson 2001).

STUDY SITES AND METHODS

We carried out 20 days of field work: seven days at North Lorma National Forest (19–25 November), eight days at Gola National Forest (27 November–3 December), and five days at Grebo National Forest (7–11 December).

At North Lorma National Forest, the habitat consisted of closed-canopy forest on flat and undulating terrain crossed by numerous small streams. The camp was situated at the intersection of a large river, the Lawa, and a tributary (at 08°01'53.6"N, 09°44' 08.6"W). The path to the camp commenced at the small village of Luyema, where agricultural plots alternated with patches of forest left standing. For the first few kilometres past the village and its immediate surroundings, the forest was degraded and open, with dense second-growth along the edge of the path which used to be a logging road. A large grassy clearing surrounded by low bushy vegetation marked the site of a former sawmill situated at the edge of closed-canopy forest. The first and last night camp was set up at Luyema, with six nights spent at the camp inside the forest.

In Gola National Forest, the main camp (07°27'09.9"N, 10°41'33.2"W) was established in closed-canopy forest on hilly terrain crossed by small, rocky streams. Traces of artisanal diamond mining were still visible. The forest was accessed from a small settlement in a large clearing, where a sawmill (now entirely destroyed) of the SLC timber company used to be (07°26'56.3"N, 10°39'05.0"W). The lateritic soil of the clearing was partly overgrown by grasses, *Chromolaena odorata* and bushes, with virtually no traces of the sawmill's structures left. The main track to the SLC clearing cut through high forest and crossed a rocky river about 1 km south of the settlement. Four nights were spent camping in the SLC clearing and four inside the forest.

In Grebo National Forest, parts of the forest had obviously been heavily logged in the relatively recent past. Patches of closed-canopy forest alternated with more open and degraded areas. Access was along a path starting at the village of Jalipo. The path had apparently been a motorable road a few years prior to our visit, but is now mostly overgrown and bordered by secondary vegetation. The first and last night, camp was set up at the edge of Jalipo (05°22'10.5"N, 07°46'14.5"W), with three nights spent at the camp inside the forest (05°24'10.4"N, 07°43'56.2"W). Before returning to Monrovia, a night was spent at the UNMIL camp in Fishtown (05°11'48.6"N, 07°52'28.8"W), allowing some bird observations to be made there on the afternoon of 11 December and the following morning.

The principal method used during this study consisted of observing birds by walking slowly along tracks and trails, if any, and stopping frequently. Notes were taken on both visual observations and bird vocalizations. Some tape-recordings were made for later deposition in sound archives. Field work was carried out from dawn (usually 06:30) until 14:00–16:00, and on a few occasions from 17:00 until sunset (around 18:30). Some species were recorded opportunistically during the night or captured in mist-nets set up for bats. Although attempts were made to cover as much ground as possible, the difficulty of access of the first two sites, North Lorma and Gola National Forests, due to the scarcity or absence of paths, meant that the areas visited in the interior of these forests were relatively limited and mostly restricted to the vicinity of the base camps. However, additional observations were made on the way to and from those camps at the start and the end of each survey.

For each field day a list was compiled of all the species that were recorded. Numbers of individuals or flocks were noted, as well as any evidence of breeding, such as the presence of juveniles, and basic information on the habitat in which the birds were observed. An attempt has been made to give indices of abundance based on the encounter rate. However, it should be noted that many bird species were not singing (e.g. cuckoos and owls) and many thus have remained unnoticed.

The weather was variable, usually overcast with some sunny spells and frequent rain, which occurred mostly at night and was occasionally heavy.

For the purposes of standardization, we have followed the nomenclature, taxonomy and sequence of Borrow and Demey (2001, 2004).

RESULTS

North Lorma National Forest

Over seven days of field work, 143 species were recorded (see Appendix 7), of which eight are of global conservation concern (BirdLife International 2000, 2004; Table 4.1). Among these, two are classified as Vulnerable (Yellow-bearded Greenbul *Criniger olivaceus* and Yellow-headed Picathartes *Picathartes gymnocephalus*), five as Near Threatened (Brown-cheeked Hornbill *Bycanistes cylindricus*, Yellow-casqued Hornbill *Ceratogymna elata*, Black-headed Rufous Warbler *Bathmocercus cerviniventris*, Rufous-winged Illadopsis *Illadopsis rufescens* and Copper-tailed Glossy Starling *Lamprotornis cupreocauda*), while one is considered Data Deficient (Yellow-footed Honeyguide *Melignomon eisentrauti*).

Seven of the 15 restricted-range species, i.e. landbird species which have a global breeding range of less than 50,000 km^2, that make up the Upper Guinea forests Endemic Bird Area (the area from Sierra Leone and south-east Guinea to south-west Ghana that encompasses the overlapping breeding ranges of restricted-range species: Stattersfield et al. 1998) were found during the study: all the above-mentioned species apart from the Yellow-casqued Hornbill *Ceratogymna elata* and the Yellow-footed Honeyguide *Melignomon eisentrauti* are of restricted range as is the non-threatened Sharpe's Apalis *Apalis sharpii* (Table 4.2). The reserve thus holds an important proportion of the Upper Guinea endemics. Of the 184 Guinea-Congo forests biome species recorded in Liberia (Robertson 2001), 97 or 53%, were recorded in North Lorma National Forest (Table 4.3). In addition, the little-known Olive Ibis *Bostrychia olivacea*, which is generally rare in Upper Guinea, was observed.

Gola National Forest

During eight days of field work, 145 species were recorded at this site (see Appendix 7), six of which are of global conservation concern (BirdLife International 2000, 2004; Table 3.1). One of these, Gola Malimbe *Malimbus ballmanni* is classified as Endangered, another, Yellow-bearded Greenbul *Criniger olivaceus*, as Vulnerable, while the remaining four are considered Near Threatened (Brown-cheeked Hornbill *Bycanistes cylindricus*, Yellow-casqued Hornbill *Ceratogymna elata*, Rufous-winged Illadopsis *Illadopsis rufescens* and Copper-tailed Glossy Starling *Lamprotornis cupreocauda*).

Six of the 15 restricted-range species that make up the Upper Guinea forests Endemic Bird Area (Stattersfield et al. 1998) were recorded from this site: all the above-mentioned species, apart from the Yellow-casqued Hornbill *Ceratogymna elata*, plus Sharpe's Apalis *Apalis sharpii* (Table 4.2).

Table 4.1. Species of global conservation concern recorded during the Liberia RAP survey.

Species	Common Name	Threat Status	Sites		
			North Lorma	Gola	Grebo
Agelastes meleagrides	White-breasted Guineafowl	VU			U
Bycanistes cylindricus	Brown-cheeked Hornbill	NT	C	C	C
Ceratogymna elata	Yellow-casqued Hornbill	NT	C	C	C
Melignomon eisentrauti	Yellow-footed Honeyguide	DD	R		
Lobotos lobatus	Western Wattled Cuckoo-shrike	VU			R
Bleda eximius	Green-tailed Bristlebill	VU			U
Criniger olivaceus	Yellow-bearded Greenbul	VU	F	U	U
Bathmocercus cerviniventris	Black-headed Rufous Warbler	NT	R		
Melaenornis annamarulae	Nimba Flycatcher	VU			F
Picathartes gymnocephalus	Yellow-headed Picathartes	VU	U		
Illadopsis rufescens	Rufous-winged Illadopsis	NT	F	R	C
Malaconotus lagdeni	Lagden's Bush-shrike	NT			R
Lamprotornis cupreocauda	Copper-tailed Glossy Starling	NT	U	C	C
Malimbus ballmanni	Gola Malimbe	EN		F	
Number of species recorded:		**14**	**8**	**6**	**10**

Threat Status (BirdLife International 2000, 2004, 2006b):
EN = Endangered: species facing a high risk of extinction in the immediate future
VU = Vulnerable: species facing a high risk of extinction in the medium-term future
NT = Near Threatened: species coming very close to qualifying as Vulnerable
DD = Data Deficient: species for which there is inadequate information to make an assessment of its risk of extinction

Encounter rate:
C – Common: encountered daily, either singly or in significant numbers
F – Fairly common: encountered on most days
U – Uncommon: irregularly encountered and not on the majority of days

Of the 184 Guinea-Congo forests biome species recorded in Liberia (Robertson 2001), 91 or 49%, were recorded in Gola National Forest.

Grebo National Forest

In five days of field work, 157 species were recorded here (see Appendix 7), of which 10 are of global conservation concern (BirdLife International 2000, 2004). Among these, five are classified as Vulnerable (White-breasted Guineafowl *Agelastes meleagrides*, Western Wattled Cuckoo-shrike *Lobotos lobatus*, Green-tailed Bristlebill *Bleda eximius*, Yellow-bearded Greenbul *Criniger olivaceus* and Nimba Flycatcher *Melaenornis annamarulae*) and five are considered Near Threatened (Brown-cheeked Hornbill *Bycanistes cylindricus*, Yellow-casqued Hornbill *Ceratogymna elata*, Rufous-winged Illadopsis *Illadopsis rufescens*, Lagden's Bush-shrike *Malaconotus lagdeni* and Copper-tailed Glossy Starling *Lamprotornis cupreocauda*).

Nine of the 15 restricted-range species that make up the Upper Guinea forests Endemic Bird Area (Stattersfield et al. 1998) were found at this site: all the above-mentioned species, apart from the Yellow-casqued Hornbill *Ceratogymna elata* and Lagden's Bush-shrike *Malaconotus lagdeni,* are of restricted range as is Sharpe's Apalis *Apalis sharpii* (Table 4.2). The site thus holds the most important proportion of the Upper Guinea endemics of the three reserves visited. Of the 184 Guinea-Congo forests biome species recorded in Liberia (Robertson 2001), 114 or 62%, were recorded in Grebo (Table 4.3). In addition, a number of rare and poorly known species were observed, including Spot-breasted Ibis *Bostrychia rara*, Congo Serpent Eagle *Urotriorchis spectabilis* and Blue-headed Bee-eater *Merops muelleri.*

NOTES ON SPECIFIC SPECIES

See Table 4.1 for explanation of threat status. Status and distribution in Liberia from Gatter (1997) and in West Africa from Borrow and Demey (2001, 2004).

Species of conservation concern

Agelastes meleagrides White-breasted Guineafowl (VU)
Two groups, numbering three and six birds respectively, were encountered at two locations in Grebo National Forest. This Upper Guinea forest endemic was formerly a not uncommon and widespread resident in Liberia, but is now rare to locally common, with its population divided in a western part (centered around Kpelle and Gola National Forests) and an extended eastern part. Together with western Côte d'Ivoire, Liberia retains the largest population of this species, which is threatened by habitat loss and hunting (BirdLife International 2006a).

Bycanistes cylindricus Brown-cheeked Hornbill (NT)
This species was common in all three forests, with daily observations of up to six individuals. This generally scarce to locally frequent Upper Guinea endemic is a not uncommon resident in high-forest blocks in Liberia.

Table 4.2. Species restricted to the Upper Guinea forests Endemic Bird Area recorded during the Liberia RAP survey.

Species	Common Name	Sites		
		North Lorma	Gola	Grebo
Agelastes meleagrides	White-breasted Guineafowl			x
Bycanistes cylindricus	Brown-cheeked Hornbill	x	x	x
Lobotos lobatus	Western Wattled Cuckoo-shrike			x
Bleda eximius	Green-tailed Bristlebill			x
Criniger olivaceus	Yellow-bearded Greenbul	x	x	x
Bathmocercus cerviniventris	Black-headed Rufous Warbler	x		
Apalis sharpii	Sharpe's Apalis	x	x	x
Melaenornis annamarulae	Nimba Flycatcher			x
Picathartes gymnocephalus	Yellow-headed Picathartes	x		
Illadopsis rufescens	Rufous-winged Illadopsis	x	x	x
Lamprotornis cupreocauda	Copper-tailed Glossy Starling	x	x	x
Malimbus ballmanni	Gola Malimbe		x	
Number of species recorded		**7**	**6**	**9**

Ceratogymna elata Yellow-casqued Hornbill (NT)
This species was also common in all three forests, with daily observations of up to six individuals. A not uncommon and widespread resident in high forest in Liberia, which is rare to uncommon and local elsewhere in its fragmented range from south-west Senegal to western Cameroon.

Melignomon eisentrauti Yellow-footed Honeyguide (DD)
One was heard singing at the edge of a mixed-species flock inside the forest at North Lorma National Forest. A rare and little-known resident throughout its range, previously recorded only from the Mt. Nimba area, Wonegizi, Mt. Balagizi and south of Vahun (Gatter 1997).

Lobotos lobatus Western Wattled Cuckoo-shrike (VU)
One was seen in a mixed-species flock at a height of c.10 m in Grebo National Forest. This Upper Guinea endemic is a locally rare to uncommon resident in Liberia, where it is known from 21 localities (Gatter 1997); it is rare elsewhere in its fragmented range.

Bleda eximius Green-tailed Bristlebill (VU)
A pair and a single were observed at two locations in Grebo National Forest. One member of the pair and the single were singing; the former was tape-recorded and reacted strongly to playback by flying towards the observer, singing continuously. Apparently a common and widespread resident in Liberia (Gatter 1997), but this Upper Guinea endemic is rare elsewhere.

Criniger olivaceus Yellow-bearded Greenbul (VU)
Fairly common at North Lorma National Forest and uncommon at Gola and Grebo National Forests. It was always observed in mixed-species flocks, usually in pairs, but six individuals were seen in a single flock at the first site. A not uncommon to locally common forest resident in Liberia (Gatter 1997), but this Upper Guinea endemic is generally rare elsewhere.

Bathmocercus cerviniventris Black-headed Rufous Warbler (NT)
A pair was found in dense vegetation near a small stream at the forest edge at Luyema, North Lorma National Forest. An uncommon to fairly common resident in Liberia (Gatter 1997); this Upper Guinea endemic has a fragmented range and is rare to uncommon and very local elsewhere.

Melaenornis annamarulae Nimba Flycatcher (VU)
Five individuals, singing from the canopy of tall trees, were found along the main track in Grebo National Forest. A rare, though probably overlooked forest resident in Liberia, where it is known from the Nimba area, Glaro and Wologizi; this Upper Guinea endemic is rare to scarce and local throughout its restricted range.

Picathartes gymnocephalus Yellow-headed Picathartes (VU)
A large rock with 20 nests in good condition was found within North Lorma National Forest. In the evening, just before dusk, five birds were observed as they came to the nesting site. A generally scarce and very local resident in the forest zone, endemic to Upper Guinea. Rare to not uncommon in Liberia; the country probably holds the largest population of this threatened species.

Illadopsis rufescens Rufous-winged Illadopsis (NT)
Recorded at all three forest sites: fairly common at North Lorma National Forest, rare at Gola National Forest and common at Grebo National Forest. At the first site a previously unknown song was tape-recorded and found to be from a member of a duetting pair, of which the other member uttered the well-known song presented by Chappuis (2000). A generally uncommon forest resident, endemic to Upper Guinea, but not uncommon to locally common and widespread in Liberia.

Malaconotus lagdeni Lagden's Bush-shrike (NT)
A single and a pair were observed in mixed-species flocks at two localities in Grebo National Forest. The single was silently foraging at c.10 m height inside the forest, but sang briefly in response to playback. The pair occurred in the canopy of tall trees at the forest edge and was singing; it was eventually seen displaying on a big horizontal branch, one member of the pair approaching the other with head held backwards while uttering a dry *krrrrrrr krrrrrrr*. This species has a disjunct distribution across West and Central Africa, with the scarce to rare nominate subspecies being endemic to Upper Guinea. Liberia's forests are estimated to hold the largest population of this taxon (BirdLife International 2004, 2006b).

Lamprotornis cupreocauda Copper-tailed Glossy Starling (NT)
Rather uncommon at North Lorma National Forest, but small numbers were observed daily at Gola and Grebo National Forests. A fairly common to locally common forest resident, endemic to Upper Guinea and widespread in Liberia.

Malimbus ballmanni Gola Malimbe (EN)
A pair with a juvenile was found in the same mixed-species flock on two consecutive days, foraging at mid-level in the interior of Gola National Forest. Another pair, also with a juvenile, was found in another mixed flock at c.1 km from the first. Gola Malimbe is a rare to locally common and rather poorly known forest resident endemic to Upper Guinea. It is known only from eastern Sierra Leone, Liberia, south-east Guinea (where discovered in Diécké Forest during a RAP in 2003), and western Côte d'Ivoire.

Other noteworthy records and range extensions
Bostrychia olivacea Olive Ibis
A single and a pair were seen on two consecutive days in North Lorma National Forest. The former flew along a forested river and eventually landed on the bank, allowing good views; the pair was flushed from a small stream in the forest

Table 4.3. Species restricted to the Guinea-Congo Forests biome recorded during the Liberia RAP survey.

Species	Common Name	Location		
		North Lorma	Gola	Grebo
Tigriornis leucolopha	White-crested Tiger Heron			x
Bostrychia rara	Spot-breasted Ibis			x
Pteronetta hartlaubii	Hartlaub's Duck			x
Dryotriorchis spectabilis	Congo Serpent Eagle			x
Urotriorchis macrourus	Long-tailed Hawk	x		
Francolinus lathami	Latham's Forest Francolin	x	x	x
Francolinus ahantensis	Ahanta Francolin		x	x
Agelastes meleagrides	White-breasted Guineafowl			x
Himantornis haematopus	Nkulengu Rail			x
Sarothrura pulchra	White-spotted Flufftail	x	x	x
Turtur brehmeri	Blue-headed Wood Dove	x	x	x
Columba iriditorques	Western Bronze-naped Pigeon	x	x	x
Columba unicincta	Afep Pigeon	x		
Psittacus erithacus	Grey Parrot	x	x	x
Agapornis swindernianus	Black-collared Lovebird			x
Tauraco macrorhynchus	Yellow-billed Turaco	x	x	x
Cercococcyx mechowi	Dusky Long-tailed Cuckoo			x
Cercococcyx olivinus	Olive Long-tailed Cuckoo	x	x	x
Centropus leucogaster	Black-throated Coucal	x	x	x
Glaucidium tephronotum	Red-chested Owlet			x
Rhaphidura sabini	Sabine's Spinetail	x	x	x
Halcyon badia	Chocolate-backed Kingfisher	x	x	x
Ceyx lecontei	African Dwarf Kingfisher	x	x	x
Alcedo leucogaster	White-bellied Kingfisher	x		x
Merops muelleri	Blue-headed Bee-eater			x
Merops gularis	Black Bee-eater		x	x
Eurystomus gularis	Blue-throated Roller	x	x	
Phoeniculus castaneiceps	Forest Wood-hoopoe	x		x
Tropicranus albocristatus	White-crested Hornbill	x	x	x
Tockus hartlaubi	Black Dwarf Hornbill			x
Tockus camurus	Red-billed Dwarf Hornbill	x	x	x
Tockus fasciatus	African Pied Hornbill	x	x	x
Bycanistes fistulator	Piping Hornbill	x	x	x
Bycanistes cylindricus	Brown-cheeked Hornbill	x	x	x
Ceratogymna atrata	Black-casqued Hornbill	x	x	x
Ceratogymna elata	Yellow-casqued Hornbill	x	x	x
Gymnobucco calvus	Naked-faced Barbet	x	x	x
Pogoniulus scolopaceus	Speckled Tinkerbird	x	x	x
Pogoniulus atroflavus	Red-rumped Tinkerbird	x	x	x

continued

Table 4.3. *continued*

Species	Common Name	Location		
		North Lorma	Gola	Grebo
Pogoniulus subsulphureus	Yellow-throated Tinkerbird	x	x	x
Buccanodon duchaillui	Yellow-spotted Barbet	x	x	x
Tricholaema hirsuta	Hairy-breasted Barbet	x	x	x
Trachylaemus purpuratus	Yellow-billed Barbet			x
Prodotiscus insignis	Cassin's Honeybird		x	x
Melignomon eisentrauti	Yellow-footed Honeyguide	x		
Melichneutes robustus	Lyre-tailed Honeyguide	x		
Indicator maculatus	Spotted Honeyguide	x		
Campethera maculosa	Little Green Woodpecker	x		x
Campethera nivosa	Buff-spotted Woodpecker	x		x
Campethera caroli	Brown-eared Woodpecker	x		x
Dendropicos gabonensis	Gabon Woodpecker	x	x	x
Dendropicos pyrrhogaster	Fire-bellied Woodpecker	x	x	x
Smithornis rufolateralis	Rufous-sided Broadbill	x		x
Psalidoprocne nitens	Square-tailed Saw-wing	x	x	x
Hirundo nigrita	White-throated Blue Swallow		x	
Lobotos lobatus	Western Wattled Cuckoo-shrike			x
Coracina azurea	Blue Cuckoo-shrike	x	x	x
Andropadus gracilis	Little Grey Greenbul	x	x	x
Andropadus ansorgei	Ansorge's Greenbul	x	x	x
Andropadus curvirostris	Cameroon Sombre Greenbul	x	x	x
Calyptocichla serina	Golden Greenbul	x	x	x
Baeopogon indicator	Honeyguide Greenbul	x	x	x
Ixonotus guttatus	Spotted Greenbul	x	x	x
Chlorocichla simplex	Simple Leaflove	x	x	x
Thescelocichla leucopleura	Swamp Palm Bulbul	x	x	x
Phyllastrephus icterinus	Icterine Greenbul	x	x	x
Bleda syndactylus	Red-tailed Bristlebill	x	x	x
Bleda eximius	Green-tailed Bristlebill			x
Bleda canicapillus	Grey-headed Bristlebill	x		x
Criniger barbatus	Western Bearded Greenbul	x	x	x
Criniger calurus	Red-tailed Greenbul	x	x	x
Criniger olivaceus	Yellow-bearded Greenbul	x	x	x
Nicator chloris	Western Nicator	x	x	x
Stiphrornis erythrothorax	Forest Robin	x	x	x
Cossypha cyanocampter	Blue-shouldered Robin Chat			x
Alethe diademata	Fire-crested Alethe	x	x	x
Neocossyphus poensis	White-tailed Ant Thrush	x	x	x
Stizorhina finschi	Finsch's Flycatcher Thrush	x	x	x

continued

Table 4.3. *continued*

Species	Common Name	Location		
		North Lorma	Gola	Grebo
Cercotrichas leucosticta	Forest Scrub Robin	x		x
Bathmocercus cerviniventris	Black-headed Rufous Warbler	x		
Apalis nigriceps	Black-capped Apalis	x		x
Apalis sharpii	Sharpe's Apalis	x	x	x
Camaroptera superciliaris	Yellow-browed Camaroptera	x	x	x
Camaroptera chloronota	Olive-green Camaroptera	x	x	x
Macrosphenus kempi	Kemp's Longbill	x	x	x
Macrosphenus concolor	Grey Longbill	x	x	x
Eremomela badiceps	Rufous-crowned Eremomela		x	x
Sylvietta virens	Green Crombec	x		x
Sylvietta denti	Lemon-bellied Crombec	x		x
Hyliota violacea	Violet-backed Hyliota		x	x
Hylia prasina	Green Hylia	x	x	x
Fraseria ocreata	Fraser's Forest Flycatcher	x	x	x
Fraseria cinerascens	White-browed Forest Flycatcher	x		
Melaenornis annamarulae	Nimba Flycatcher			x
Muscicapa cassini	Cassin's Flycatcher	x	x	
Muscicapa olivascens	Olivaceous Flycatcher			x
Muscicapa ussheri	Ussher's Flycatcher		x	x
Myioparus griseigularis	Grey-throated Flycatcher		x	x
Erythrocercus mccallii	Chestnut-capped Flycatcher		x	x
Elminia nigromitrata	Dusky Crested Flycatcher	x		
Trochocercus nitens	Blue-headed Crested Flycatcher	x	x	x
Terpsiphone rufiventer	Red-bellied Paradise Flycatcher	x	x	x
Megabyas flammulatus	Shrike Flycatcher	x	x	x
Dyaphorophyia castanea	Chestnut Wattle-eye	x	x	x
Dyaphorophyia blissetti	Red-cheeked Wattle-eye	x		x
Batis poensis	Bioko Batis		x	
Picathartes gymnocephalus	Yellow-headed Picathartes	x		
Illadopsis fulvescens	Brown Illadopsis	x		x
Illadopsis cleaveri	Blackcap Illadopsis	x		x
Illadopsis rufescens	Rufous-winged Illadopsis	x	x	x
Pholidornis rushiae	Tit-hylia		x	x
Anthreptes gabonicus	Brown Sunbird	x		
Anthreptes rectirostris	Green Sunbird		x	x
Anthreptes seimundi	Little Green Sunbird	x		
Deleornis fraseri	Fraser's Sunbird	x	x	x
Cyanomitra cyanolaema	Blue-throated Brown Sunbird	x	x	x
Chalcomitra adelberti	Buff-throated Sunbird		x	x

continued

Table 4.3. *continued*

Species	Common Name	Location		
		North Lorma	Gola	Grebo
Cinnyris johannae	Johanna's Sunbird		x	x
Cinnyris superbus	Superb Sunbird		x	
Dryoscopus sabini	Sabine's Puffback		x	x
Laniarius leucorhynchus	Sooty Boubou	x		
Prionops caniceps	Red-billed Helmet-shrike		x	x
Oriolus brachyrhynchus	Western Black-headed Oriole	x	x	x
Dicrurus atripennis	Shining Drongo	x	x	x
Onychognathus fulgidus	Forest Chestnut-winged Starling	x		x
Lamprotornis cupreocauda	Copper-tailed Glossy Starling	x	x	x
Malimbus ballmanni	Gola Malimbe		x	
Malimbus scutatus	Red-vented Malimbe	x	x	x
Malimbus malimbicus	Crested Malimbe	x	x	x
Malimbus nitens	Blue-billed Malimbe	x	x	x
Malimbus rubricollis	Red-headed Malimbe			x
Ploceus nigerrimus	Vieillot's Black Weaver	x		x
Ploceus albinucha	Maxwell's Black Weaver		x	
Nigrita bicolor	Chestnut-breasted Negrofinch	x	x	x
Spermophaga haematina	Western Bluebill		x	x
Pyrenestes sanguineus	Crimson Seedcracker		x	
Number of species recorded		**97**	**90**	**116**

interior. All birds were silent. A not uncommon resident in large forest blocks in Liberia, but not mentioned for the Wologizi area by Gatter (1997).

Bostrychia rara Spot-breasted Ibis
One flew over the village of Jalipo, Grebo National Forest, at dawn, calling loudly, on 7 and 11 December. A rare resident in Liberia and elsewhere in Upper Guinea.

Ciconia episcopus Woolly-necked Stork
One flew over the main track to the SLC clearing, Gola National Forest, on 3 December. A new locality for this resident and not uncommon passage migrant.

Pteronetta hartlaubii Hartlaub's Duck
A female and a pair were observed at two sites in Grebo National Forest. A not uncommon and widespread resident in Liberia. Formerly considered Near Threatened (BirdLife International 2000, 2004), but recently downlisted to Least Concern because of its large range and global population size. In West Africa, however, it seems to have suffered major declines and is now very scarce, with perhaps fewer than 1,000 individuals remaining (BirdLife International 2006c).

Dryotriorchis spectabilis Congo Serpent Eagle
One landed on a perch inside the forest near the main track to Jalipo village, Grebo National Forest, on 11 December. The adult plumage still had some traces of immaturity (a few white tips to the feathers of the small crest, a few pale-tipped wing-coverts, underparts pure white heavily blotched dark brown without barring on the flanks). A not uncommon, but rarely seen resident of high and old secondary forest in Liberia.

Accipiter melanoleucus Black Sparrowhawk
An adult of the typical form and a melanistic individual were seen in the SLC clearing, Gola National Forest. A not uncommon forest resident; widespread in Liberia, but not previously recorded in Gola National Forest.

Poicephalus gulielmi Red-fronted Parrot
A group of six silently flew over the village of Jalipo, Grebo National Forest, on 11 December. Birds in the westernmost part of the species' range, from Liberia to Ghana, are from the race *fantiensis*, which is generally rare to uncommon. In Liberia, it is an uncommon and local resident, known only from the east.

Agapornis swindernianus Black-collared Lovebird
Observed at two locations in Grebo National Forest, on 8 December (one individual) and the next day (two together). The nominate subspecies is a rare Upper Guinea endemic and Gatter (1997) mentions that it is a very rare or extinct resident in Liberia.

Glaucidium tephronotum Red-chested Owlet
Singles heard calling at two localities in Grebo National Forest, at night and during the daytime on 9–10 December, were the only owls heard during the entire survey. The nominate subspecies is an Upper Guinea endemic and a rare to not uncommon forest resident in Liberia.

Merops muelleri Blue-headed Bee-eater
A pair was observed along the main track near the village of Jalipo, Grebo National Forest, on 11 December. In West Africa, this species is generally a scarce to rare and local forest resident; in Liberia it is uncommon but widespread.

Phoeniculus castaneiceps Forest Wood-hoopoe
One record each for North Lorma (one individual) and Grebo (two) National Forests. A generally scarce to uncommon and local forest resident in West Africa, but not uncommon in Liberia. Care should be taken when mapping this species on call alone, as at both localities we observed Shining Drongo *Dicrurus atripennis* uttering an almost perfect imitation of *P. castaneiceps*' call.

Prodotiscus insignis Cassin's Honeybird
One at Gola National Forest and another at Grebo National Forest. New localities. A rare to uncommon resident in Liberia.

Dendropicos gabonensis Gabon Woodpecker
Recorded at all three sites. Although this is a locally not uncommon to common resident in Liberia, it is considered rare in forest blocks by Gatter (1997), who does not map it in Gola or Grebo National Forests.

Cossypha cyanocampter Blue-shouldered Robin Chat
A singing bird was observed near Jalipo village, Grebo National Forest, on 11 December (with another at Fishtown the next day). Not mapped for the area by Gatter (1997).

Cercotrichas leucosticta Forest Scrub Robin
Recorded at North Lorma and Grebo National Forests. Although mapped for these areas by Gatter (1997), there seem to be few records.

Turdus pelios African Thrush
Although stated to be confined to coastal and northern savannas and not entering clearings in the large forest blocks (Gatter 1997), we recorded singles in the SLC clearing, Gola National Forest, on 27 November and 2 December.

Hippolais (pallida) opaca Western Olivaceous Warbler
Up to three were observed in detail as they foraged at the edge of a patch of *Chromolaema odorata* in the SLC clearing, Gola National Forest, on 27 November–3 December. New locality. Only two previous records of this Palearctic migrant mentioned by Gatter (1997).

Cisticola brachypterus Short-winged Cisticola
A singing pair in the SLC clearing, Gola National Forest, on 27 November–3 December. New locality. Said to occur only in coastal and northern savannas by Gatter (1997).

Apalis nigriceps Black-capped Apalis
Common in Grebo National Forest, with daily records of up to four singing birds. New locality; not mapped for the east by Gatter (1997). Also recorded in North Lorma National Forest (once a single singing bird). According to Gatter (1997), a rare to not uncommon resident in northern highlands, above 500 m, but our observations confirm that this species also occurs in lowland forest.

Apalis sharpii Sharpe's Apalis
This Upper Guinea endemic was common at all three sites, with daily observations of up to seven birds.

Myioparus griseigularis Grey-throated Flycatcher
One singing bird was seen and tape-recorded at Gola National Forest on 2 December and singles were found at two locations in Grebo National Forest, on 7 and 9 December. New localities; previously only recorded from Yekapa/Nimba and considered a rare, though probably overlooked, resident (Gatter 1997).

Myioparus plumbeus Lead-coloured Flycatcher
One was singing at Luyema and another at a clearing 3 km further, North Lorma National Forest, on 19 November, and one was heard and seen well at the SLC clearing, Gola National Forest, on 27 November–3 December. The only records mentioned by Gatter (1997) are two collected in 1891 near Monrovia.

Cinnyricinclus leucogaster Violet-backed Starling
Two females in the SLC clearing, Gola National Forest, on 1 December. New locality. A common dry season visitor to coastal savannas, but rare inland (Gatter 1997).

Evidence of breeding

Stiphrornis erythrothorax Forest Robin
An immature was mist-netted in Grebo National Forest on 9 December.

Apalis sharpii Sharpe's Apalis
A pair with a juvenile along the road to the SLC clearing on 1 December.

Hylia prasina Green Hylia
A pair with a juvenile in Grebo National Forest on 10 December.

Dyaphorophyia castanea Chestnut Wattle-eye
Parents accompanied by a juvenile were seen at two sites in Grebo National Forest, on 7 and 10 December.

Batis poensis Bioko Batis
A pair was observed at its nest in Gola National Forest on 27 November. The moss-covered nest was at 20 m height on a bare horizontal branch of a tall tree next to a wide forest track. On 3 December, it was seen to contain two feathered nestlings. New locality for this rare, but probably overlooked forest resident in Liberia and the third nest record for its entire range (Urban et al. 1997).

Spermophaga haematina Western Bluebill
A few juveniles with an adult in Gola National Forest on 1 December.

Pyrenestes sanguineus Crimson Seedcracker
Several juveniles with adults in the SLC clearing, Gola National Forest, 27 November – 3 December.

Miscellaneous noteworthy records made outside the three surveyed forests

A few records, made outside the surveyed forests, are worthy of note, as they constitute new localities for the species involved, based on the species accounts and distribution maps in Gatter (1997).

Falco biarmicus Lanner Falcon
One seen at Monrovia on 14 December (with O. Langrand, F. Molubah and K.-D. Dijkstra). Only mapped for the north of the country by Gatter (1997), where said to be a dry-season visitor.

Dendropicos fuscescens Cardinal Woodpecker
A male seen well on one of the few trees in the UNMIL compound at Voinjama (08°25'N, 09°45'W) on 26 November. Only one record, from 1984 near Bawomai (08°28'N, 09°55'W), mentioned by Gatter (1997).

Psalidoprocne obscura Fanti Saw-wing
Up to 25 seen at Fishtown on 11–12 December. This African migrant, said to be rare to not uncommon in Liberia, is not mapped for the area by Gatter (1997).

Hirundo abyssinica Lesser Striped Swallow
Up to four seen at Fishtown on 11–12 December. Not mapped for the south-east by Gatter (1997).

Hirundo preussi Preuss's Cliff Swallow
A colony of c.100 active nests under the eaves of the UNMIL headquarters at Voinjama. Said to be a rare (dry season?) visitor by Gatter (1997).

Anthus leucophrys Plain-backed Pipit
A pair, one member of which was singing, on farmland at Fishtown on 12 December. Not mapped for the area by Gatter (1997).

Oriolus nigripennis Black-winged Oriole
One at Fishtown on 12 December. Not mapped for the area by Gatter (1997).

DISCUSSION

The total number of 211 species recorded across all three sites, representing about a third of Liberia's avifauna, is relatively high considering the short study period (20 days of field work) and the limited area that could be covered due to the difficulty of access to two of the sites. Several records represent range extensions, as compared to the distribution maps in Gatter (1997) and Borrow and Demey (2004). New and reliable information on the avifauna of the three forests and on the distribution of birds species within Liberia has thus been gathered. By comparison, 179 species were recorded in Haute Dodo and Cavally Forest Reserves, Côte d'Ivoire, over 15 days in 2002 (Demey and Rainey 2005) and 170 in Draw River, Boi Tano and Krokosua Hills Forest Reserves, Ghana, over 17 days in 2003 (Rainey and Asamoah 2005).

Twelve of the 15 restricted-range species, i.e. species which have a global breeding range of less than 50,000 km^2, that make up the Upper Guinea forests Endemic Bird Area and 136 of the 184 Guinea-Congo forests biome species recorded in Liberia (Stattersfield et al. 1998, Robertson 2001), or 74%, were found during the study. Several of the biome-restricted species had not been recorded for the IBAs in which the three forests are partly or entirely included (Robertson 2001): four in North Lorma National Forest, 24 in Gola National Forest, and 11 in Grebo National Forest (Table 4.3).

For North Lorma National Forest, the good condition of the forest and the records of the poorly-known Yellow-footed Honeyguide and the colony of the charismatic Yellow-headed Picathartes are particularly noteworthy.

Although only six species of global conservation concern were found in Gola National Forest, one of these was the Gola Malimbe, the rarest and most threatened species on our list. This is especially noteworthy as access to this forest was particularly difficult and field work inside the forest proper was limited to three days in a small area in the camp environs.

Despite the heavy logging to which parts of Grebo National Forest were subjected in the past, the site still contained substantial patches of good forest, as attested by the presence of White-breasted Guineafowl, which was recorded at this site only. Although only a small part of Grebo National Forest, east of our study site, was selected by Robertson (2001) to form the Cavalla River IBA, presumably because that area was proposed as a Nature Reserve in 1983, our field work suggests that a much larger part, possibly even all of the National Forest, qualifies as an IBA. We have therefore compared our Grebo National Forest species list given in Table 4.3 with the list given for the more restricted Cavalla River IBA by Robertson (2001).

Overall, the rapid survey of the three National Forests was definitely successful in its aim to collect fresh and reliable data on their avifauna and thus produce a rapid, first-cut assessment of the value for birds of these relatively poorly-known sites. The species accumulation rates suggest that the number of species will continue to rise substantially after further survey work.

CONSERVATION RECOMMENDATIONS

Considering the high conservation value of the three forests, the following recommendations are made:

Carry out further surveys at different times of year to complete the avifaunal species lists, evaluate the threats to species of conservation concern and estimate their population sizes.

Associate Liberian conservation NGOs and local villagers with the survey work, to strengthen capacity of the former and generate awareness of conservation issues in the latter.

Keep substantial portions of all three forests free from logging, in order to aid survival of plant and animal species that require intact high forest and permit recolonisation of logged areas. Although many forest birds may survive in logged forest for some time at least, certain species, such as White-breasted Guineafowl, need an open understorey to survive in the long term; this kind of habitat can only be found in intact forest. Although Grebo National Forest had been logged fairly recently, the alternation of closed-canopy forest patches with degraded areas made for a high bird diversity with species dependent on good forest also venturing into logged areas.

Rigorously control the awarding of logging concessions and the logging process itself. The type of logging can indeed make a crucial difference to the ultimate survival of forest-restricted species. Carefully implemented selective logging, causing a limited amount of collateral damage and leaving a significant percentage of 'uneconomic' tree species standing, may allow certain bird species to remain in the area and also favor forest regeneration. The majority of the species of global concervation concern recorded during this study occurred within or at the edge of high forest and were absent from the more degraded areas that had been extensively logged.

Curtail hunting within the three forests. Although hunters currently mainly target mammals, certain large bird species, such as White-breasted Guineafowl, Crested Guineafowl, Great Blue Turaco and large hornbills, also fall victim to hunting, which could constitute a major threat to the survival of these species.

Include the whole or at least the major part of Grebo National Forest into an enlarged Cavalla River IBA, as our data suggest that the site holds both significant numbers of globally threatened species and a significant component of species restricted to the Guinea-Congo forests biome.

Examine the feasability to confer National Park status to Grebo National Forest in order to raise the protection status of the site. As stated before our data suggests that there are significant numbers of species of conservation concern that are restricted to these forests.

Put in place monitoring programs to assess the impact of all human activities on bird populations and associate local communities (especially hunters) with these.

Publicize the organisation of conservation-related actions in the three forests and make all information concerning these sites widely available.

REFERENCES

Allport, G., M. Ausden, P.V. Hayman, P. Robertson and P.Wood. 1989. The Conservation of the Birds of Gola Forest, Sierra Leone. Study Report No. 38. International Council for Bird Preservation, Cambridge, UK.

BirdLife International. 2000. Threatened Birds of the World. Lynx Edicions and BirdLife International. Barcelona, Spain and Cambridge, UK.

BirdLife International. 2004. Threatened Birds of the World 2004. CD-ROM. BirdLife International. Cambridge, UK.

BirdLife International. 2006a. Species factsheet: *Agelastes meleagrides.* Downloaded from http://www.birdlife.org on 3 August 2006.

BirdLife International. 2006b. Species factsheet: *Malaconotus lagdeni.* Downloaded from http://www.birdlife.org on 3 August 2006.

BirdLife International. 2006c. Species factsheet: *Pteronetta hartlaubii.* Downloaded from http://www.birdlife.org on 3 August 2006.

Borrow, N. and R. Demey. 2001. Birds of Western Africa. Christopher Helm. London.

Borrow, N. and R. Demey. 2004. Field Guide to the Birds of Western Africa. Christopher Helm. London.

Chappuis, C. 2000. African Bird Sounds: Birds of North, West and Central Africa and Neighbouring Atlantic Islands. 15 CDs. Société d'Etudes Ornithologiques de France and British Library National Sound Archive. Paris and London.

Demey, R. and H.J. Rainey. 2004. A preliminary survey of the birds of the Forêt Classée du Pic de Fon. *In*: McCullough, J. (ed.). A Rapid Biological Assessment of the Forêt Classée du Pic de Fon, Simandou Range, South-eastern Republic of Guinea. RAP Bulletin of Biological Assessment 35. Conservation International. Washington, DC. Pp. 63–68.

Demey, R. and H.J. Rainey. 2005. A rapid survey of the birds of Haute Dodo and Cavally Classified Forests. *In*: Alonso, L. E., F. Lauginie and G. Rondeau (eds.). A Rapid Biological Assessment of Two Classified Forests in South-western Côte d'Ivoire. RAP Bulletin of Biological Assessment 34. Conservation International. Washington, DC. Pp. 84–90.

Gatter, W. 1997. Birds of Liberia. Pica Press. Robertsbridge.

Gartshore, M.E. 1989. An Avifaunal Survey of Tai National Park, Ivory Coast. Study Report No. 39. International Council for Bird Preservation, Cambridge, UK.

Gartshore, M.E., P.D. Taylor and I.S. Francis. 1995. Forest Birds in Côte d'Ivoire. A survey of Tai National Park and other forests and forestry plantations, 1989–1991. Study Report No. 58. BirdLife International, Cambridge, UK.

ICBP. 1992. Putting Biodiversity on the Map: Priority Areas for Global Conservation. International Council for Bird Preservation. Cambridge, UK.

Mittermeier, R.A., P. Robles Gil, M. Hoffmann, J. Pilgrom, T. Brooks, C.G. Mittermeier, J. Lamoreux and G.A.B. da Fonseca (eds.). 2004. Hotspots Revisited. Earth's Biologically Richest and Most Endangered Terrestrial Ecoregions. CEMEX/Agrupación Sierra Madre, Mexico City.

Rainey, H.J. and A. Asamoah. 2005. Rapid assessment of the birds of Draw River, Boi-Tano and Krokosua Hills. *In*: McCullough, J., J. Decher and D.G. Kpelle (eds.). A Biological Assessment of the Terrestrial Ecosystems of the Draw River, Boi-Tano, Tano Nimiri and Krokosua Hills Forest Reserves, Southwestern Ghana. RAP Bulletin of Biological Assessment 36. Conservation International. Washington, DC. Pp. 50–56.

Robertson, P. 2001. Liberia. *In:* L.D.C. Fishpool and M.I. Evans (eds.). Important Bird Areas in Africa and Associated Islands: Priority Sites for Conservation. Pisces Publications and BirdLife International, Newbury and Cambridge, UK. Pp. 473–480.

Stattersfield, A.J, M.J. Crosby, A.J. Long and D.C. Wege. 1998. Endemic Bird Areas of the World: Priorities for Biodiversity Conservation. BirdLife International. Cambridge, UK.

Urban, E.K., C.H. Fry and S. Keith (eds.). 1997. The Birds of Africa. Vol. 5. Academic Press, London.

Chapter 5

Rapid survey of bats of North Lorma, Gola and Grebo National Forests, with notes on shrews and rodents

Ara Monadjem and Jakob Fahr

SUMMARY

Bats were sampled in three National Forests in Liberia using mist nets, a harp trap, and roost searches. Terrestrial small mammals were captured opportunistically and were not used in the final assessment of the forests. A total of 182 bats of 22 species was captured, representing 37% of the bat species known to occur in Liberia. Species richness was highest at Gola and Grebo National Forests, possibly because secondary forest and forest edge was sampled there. North Lorma National Forest, where only forest interior was surveyed, had both the lowest capture success and the lowest species richness. Three IUCN Red List species were recorded: *Rhinolophus hillorum* (Vulnerable) in Gola National Forest, *Scotonycteris zenkeri* (Near Threatened) in Grebo National Forest, and *Hipposideros fuliginosus* (Near Threatened) in North Lorma National Forest. Bat assemblages in each of the surveyed areas were characterized by forest-dependent species. Not a single species typical of savanna habitats was recorded, indicating high habitat integrity of the National Forests. Three species are reported for the first time from Liberia (*Rhinolophus landeri, Neoromicia guineensis, Neoromicia* aff. *grandidieri*), raising the species total for the country to 59. An updated checklist with corrected species identifications is presented for the bats of Liberia. Two species of shrews, one murid rodent, five squirrels and one anomalure (scaly-tailed squirrel) were also recorded, including the rarely reported Western Palm Squirrel *Epixerus ebii* and the Lesser Anomalure *Anomalurus* cf. *pusillus.*

INTRODUCTION

The Upper Guinean forest region between Guinea in the west and Togo in the east has been recognized as a global biodiversity hotspot (Myers et al. 2000, Bakarr et al. 2004, Küper et al. 2004). However, only 15–20% of relatively undisturbed forest is thought to remain today, of which 44% lies in Liberia (Päivinen et al. 1992, Bakarr et al. 2004). Liberia is the only West African country entirely situated within the forest zone and still harbors 4.4 million hectares of forest (about 46% of the country's land area), of which 2.4 million hectares are considered as relatively pristine (Bayol and Chavalier 2004). The annual deforestation rate in Liberia was estimated at 1.6% between 1990 and 2000 and 1.8% between 2000 and 2005 (FAO 2006). The rapid destruction of Liberia's forest resources during the last 15 years has been caused by over-harvesting and uncontrolled logging, which was particularly aggravated by civil war (ITTO 2006). At a priority-setting workshop for biodiversity conservation in the Upper Guinean forest region, Liberia was identified as the highest-priority country with over 35% of the country ranked as "Exceptionally High" priority (Bakarr et al. 2001). Liberia indeed supports extensive forest cover providing habitat for numerous threatened plants and animals. Despite this, it remains one of the most poorly surveyed countries within the Upper Guinean forest region and new distribution data are needed to assess the state of the remaining forest blocks.

A rapid survey was therefore carried out in three National Forests, North Lorma, Gola and Grebo, in order to provide the Government of Liberia with relevant biological information to

Table 5.1: Coordinates and habitat of the five sites of the RAP survey where bats were sampled.

Site	Location	Coordinates	Habitat
Lorma	North Lorma, camp site	8°01'54"N, 09°44'09"W	forest interior
Gola 1	Gola, camp site	7°27'10"N, 10°41'33"W	forest interior
Gola 2	Gola, S.L.C. village	7°26'56"N, 10°39'05"W	secondary forest and forest edge
Grebo 1	Grebo, camp site	5°24'10"N, 07°43'56"W	forest interior
Grebo 2	Grebo, Jalipo village	5°22'11"N, 07°46'15"W	secondary forest and forest edge

rank these areas according to their relevance for biodiversity conservation. Conservation planning traditionally relies on species distributions with the aim of delineating and prioritizing networks of protected areas (Howard et al. 2000).

This study reports on the small mammals recorded in these forests. Small mammals traditionally include shrews (Soricomorpha), rodents (Rodentia) and bats (Chiroptera). Due to logistical problems and time constraints, trapping of terrestrial small mammals (shrews and rodents) was only conducted intermittently at the three forests. This report focuses on data pertaining to bats, with brief notes on terrestrial small mammals. Bats are a particularly suitable group for setting conservation priorities due to their high diversity (they are usually the most species-rich mammalian order in tropical communities), pronounced species-specific habitat requirements and patterns of endemism (many species have small distribution ranges). Furthermore, bats provide crucial ecosystem services as pollinators and seed dispersers of plants as well as predators of insects.

The majority of historic bat records from Liberia originated from the lowland forests near the coast (Jentink 1888, Miller 1900, Allen and Coolidge 1930, Kuhn 1962, 1965). More recently, intensive surveys and ecological studies focused on the area around Mt. Nimba in the northeast of the country (Coe 1975, Verschuren 1977, Hill 1982, Wolton et al. 1982). The bats of Liberia were reviewed by Koopman (1989) and Koopman et al. (1995), who listed 57 certain and five questionable species for the country. The latter studies included new species records from all over the country, with particularly interesting results from the northwest (Voinjama area and Wonegizi Mts.). Prior to the present study, North Lorma, Gola and Grebo National Forests had never been surveyed for bats.

METHODS

Study sites

North Lorma National Forest is situated in northwestern Liberia near the border with Guinea and constitutes an important forest corridor between the Wologizi and the Wonegizi Mountains. These two mountain ranges, which include Liberia's highest peak, Mt. Wutewe (1424 m), form the most important montane region in Liberia apart from Mt. Nimba. Annual precipitation at North Lorma National Forest is approximately 2500 mm and the annual mean temperature is 24.9°C (Hijmans et al. 2005; Figure 5.1). Gola National Forest is situated between the Gola Strict Nature Reserve in Sierra Leone and Kpelle National Forest in Liberia. These reserves form a very large, contiguous forest tract and offer the possibility for a transfrontier park. Annual precipitation at Gola National Forest is approximately 2700 mm and the annual mean temperature is 25.4°C (Hijmans et al. 2005; Figure 5.1). Grebo National Forest is situated in the southeast of the country and is contiguous with the Forêt Classée du Cavally in Côte d'Ivoire. Taï National Park in Côte d'Ivoire is very close to Grebo National Forest, but forest habitat is broken by a narrow strip of dense human settlement and farming on the Ivorian side. Annual precipitation at Grebo National Forest is approximately 2500 mm and the annual mean temperature is 25.7°C (Hijmans et al. 2005; Figure 5.1). Grebo National Forest is part of a southeastern block of wet evergreen forest, whereas North Lorma and Gola National Forests are part of the more seasonal moist evergreen and semi-deciduous forests in the northwest of the country.

Table 5.2: Number of bat species per family sampled during the RAP survey and percentage of the total species richness recorded for Liberia (see Appendix 7).

Family	Number of species	Percentage of known Liberian bat fauna
Pteropodidae	7	63.4
Emballonuridae	0	–
Nycteridae	1	16.7
Rhinolophidae	2	33.3
Hipposideridae	5	63.0
Vespertilionidae	7	35.0
Molossidae	0	–
Total	**22**	**35.0**

Sampling and data analysis

Three National Forests were surveyed by AM for five or six nights at the end of the wet season (November–December) of 2005: North Lorma National Forest from 19–24 November, Gola National Forest from 28 November – 2 December, and Grebo National Forest from 6–11 December. One site was sampled at North Lorma National Forest, whereas two sites were sampled in each of Gola and Grebo National Forests. The location of each site was recorded with a GPS receiver using WGS 84 datum (Garmin eTrex; see Table 5.1). A gazetteer of localities sampled and those discussed is presented in Appendix 8.

Standard survey techniques were employed for bats and terrestrial small mammals (Voss and Emmons 1996, Martin et al. 2001). Bats were sampled with 6, 10, 15 and 18 m mist nets near ground level, which were opportunistically set across presumed flyways such as rivers, gaps in the forest, and around fruiting trees. At each site, an elevated net was installed that consisted of two stacked 15 or 18 m nets raised 6–8 m above the ground. One two-bank harp trap (Bat Conservation and Management, model "G4 Forest Strainer," catch area 3.9 m^2) was also employed at each site. Nets and the harp trap were opened before sunset, around 18:00 hrs, and checked every 30–45 minutes until about 23:00 hrs. Nets were checked a last time at 06:00 hrs, after which they were closed. Capture success is expressed as number of individual bats caught per net hour (calculated as 12 m-net equivalents; Table 5.3).

Forty-nine voucher specimens of bats were collected and preserved in 70% ethanol. These specimens are deposited

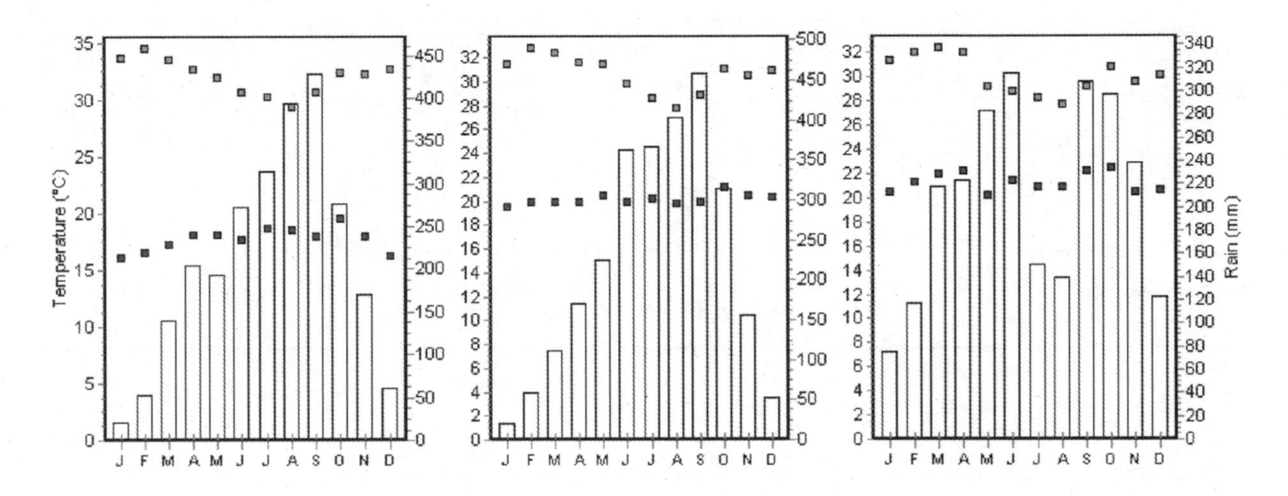

Figure 5.1: Climate diagrams, from left to right, for North Lorma, Gola, and Grebo National Forests (upper line of squares: mean monthly maximum temperature, lower line of squares: mean monthly minimum temperature, bars: mean monthly precipitation; plotted with DIVA-GIS, data from Hijmans et al. 2005).

Table 5.3: Capture effort (nh: total net hours per site, calculated as 12 m-net equivalents; th: trap hours), capture success (number of individuals; bats per net/trap hour) and species total of the RAP survey. Bats: all individuals; Mega: fruit bats only; Micro: insect bats only. Note that specimens taken in day roosts are included in the species total but not in the capture success (N° of Indiv.; Bats / nh or th).

Site	Mist Nets					Harp Trap			Roosts	Species Total
	Effort [nh]	N° of Indiv.	Bats / nh	Mega / nh	Micro / nh	Effort [th]	N° of Indiv.	Bats / th	N° of Indiv.	
Lorma	573	9	0.02	0.01	0.01	60	2	0.03	4	7
Gola 1	225	13	0.06	–	0.06	24	3	0.13	–	2
Gola 2	231	60	0.26	0.23	0.03	24	3	0.13	2	12
Grebo 1	303	35	0.12	0.11	0.01	36	0	–	3	7
Grebo 2	198	46	0.23	0.18	0.05	–	–	–	2	11
All sites	**1530**	**163**	**0.11**	**0.08**	**0.02**	**144**	**8**	**0.06**	**11**	**22**

in the research collection of JF (Department of Experimental Ecology, University of Ulm) and in the Zoologisches Forschungsmuseum Alexander Koenig, Bonn (ZFMK). The latter institution also houses the shrews and murids collected during the survey (Appendix 9). Tissue samples (wing punches) were taken from most of the collected bats and preserved in 99% ethanol. Echolocation calls of hand-held microbats were recorded with a Pettersson D240x bat detector and transferred to a Sony Walkman Professional WM-D6C. Due to operating problems, the time-expanded calls were not captured on tape and are lost for analysis. Comparative bat specimens were examined by JF from the following institutions: American Museum of Natural History, New York (AMNH); Carnegie Museum of Natural History, Pittsburgh (CM); Staatliches Museum für Naturkunde Stuttgart (SMNS); National Museum of Natural History, Smithsonian Institution, Washington, DC (USNM).

Table 5.4: Bat species recorded at each site (numbers refer to captured individuals, parentheses indicate those from day roosts). Red List: international Red List status (VU: Vulnerable, NT: Near Threatened, n.a.: not assessed; IUCN 2006). Habitat: coarse assignment to preferred habitat type (F: forest; S: savannas and woodlands; in parentheses: marginally used habitat type). For taxonomic remarks, see Appendix 10.

Species	Sites					Total	Red List	Habitat	
	North Lorma	Gola 1	Gola 2	Grebo 1	Grebo 2				
Pteropodidae									
Epomops buettikoferi			19	14	23	56		F	(S)
Hypsignathus monstrosus			8	3	2	13		F	(S)
Nanonycteris veldkampii	5		6	2	2	15		F	(S)
Scotonycteris zenkeri					1	1	NT	F	
Megaloglossus woermanni			4	1	2	7		F	
Myonycteris torquata	1		13	13	6	33		F	(S)
Rousettus aegyptiacus	(1)		4			5		F	S
Nycteridae									
Nycteris arge	(2)					2		F	
Rhinolophidae									
Rhinolophus hillorum		2	1			3	VU	F	
Rhinolophus alcyone				1		1		F	(S)
Hipposideridae									
Hipposideros ruber	(1)	14				15		F	(S)
Hipposideros fuliginosus	3					3	NT	F	
Hipposideros beatus	2			1(3)	1	7		F	
Hipposideros cyclops			2			2		F	
Hipposideros gigas			1			1		F	
Vespertilionidae									
Myotis bocagii			(2)			2		F	S
Hypsugo (crassulus) bellieri			3			3	n.a.	F	
Neoromicia nanus					6(2)	8		F	S
Neoromicia guineensis					1	1		(F)	S
Neoromicia aff. *grandidieri*			2			2	n.a.	F	
Neoromicia tenuipinnis					1	1		F	(S)
Glauconycteris poensis					1	1		F	
Specimens total	**15**	**16**	**65**	**38**	**48**	**182**			
Species total		**2**	**12**	**7**	**11**				
	7	**13**		**12**		**22**			

At each site, traplines were set for 1–3 nights. Traplines were chosen to cover as many different types of microhabitats as possible and consisted of 40 Sherman live traps, approximately 50 Museum Special snap traps and eight rat traps. Trap stations were 5–10 m apart. Traps were baited with either palm nut or a mixture of peanut butter and oats. Shrews were also captured in pitfall traps set by the herpetological team (see Chapter 3).

Smoothed species accumulation curves were generated for bat captures with the program EstimateS, Version 7.5 (Colwell 2005). These sample-based rarefaction curves were calculated with the "Mao Tau"-function (see Colwell et al. 2004) and graphs were rescaled by individuals to allow for comparison of species richness. Although there are several statistical methods for estimating the total species number from samples (e.g. Colwell 2005), these were not employed because sampling effort and methods varied greatly between nights and sites. The IUCN Red List status is based on the recent update that followed the Global Mammal Assessment (GMA) of African small mammals in January 2004 (IUCN 2006). If not otherwise stated, taxonomy follows Wilson and Reeder (2005).

RESULTS

Bats

A total of 182 individuals of 22 species and five families were captured. The families with the highest species richness were the fruit bats (Pteropodidae) and the vespertilionids (Vespertilionidae). The seven species of fruit bats captured during this study represent two-thirds of the fruit bat species known to occur in Liberia. Species coverage for the other families ranged between 17 and 63%, and two families (Emballonuridae and Molossidae) were not recorded at all during this survey (Table 5.2).

Most of the bats were captured in mist nets (Table 5.3), while lower numbers were caught with the harp trap and from day roosts. Capture success with mist nets ranged from 0.02 to 0.26 bats per net hour (12 m-equivalent). Capture success was low in the forest interior (Lorma, Gola 1 and Grebo 1), and high in secondary forest and forest edge (Gola 2 and Grebo 2). Mean capture success was 0.11 bats per net hour. Capture success of the harp trap was relatively low (0.03–0.13 bats per trap hour), but one species (*Hypsugo* [*crassulus*] *bellieri*) was recorded exclusively with this method. Roost searches further complemented the species inventory: two species (*Nycteris arge* and *Myotis bocagii*) were recorded only with this method. A cave in North Lorma National Forest harbored a colony of over a thousand *Rousettus aegyptiacus* and a small colony of about 20 *Hipposideros ruber*. *Nycteris arge* was found in a hollow tree trunk, *Hipposideros beatus* in the hollow trunk of a fallen tree, *Neoromicia nanus* in the thatched roof of a traditional house in Jalipo village, and *Myotis bocagii* in a furled banana leaf.

The most frequently captured species were the fruit bats *Epomops buettikoferi* and *Myonycteris torquata*, which accounted for 49% of all captures. In contrast, six species were captured only once and four species only twice (Table 5.4). A total of 13 species was recorded from Gola National Forest, 12 from Grebo National Forest, and seven from North Lorma National Forest. Of the seven species captured at North Lorma National Forest, *Nycteris arge* and *Hipposideros fuliginosus* were not recorded from the other two forests. Six of the 13 species captured at Gola National Forest, and six of the 12 species from Grebo National Forest were not found elsewhere (Table 5.4). Three globally threatened species (IUCN 2006) were recorded during this survey: *Rhinolophus hillorum* (Vulnerable) at Gola National Forest, *Scotonycteris zenkeri* (Near Threatened) at Grebo National Forest, and *Hipposideros fuliginosus* (Near Threatened) at North Lorma National Forest.

Of the 22 species recorded during this study, 18 are categorized as forest species (11 species restricted to forest, seven marginally extending into savanna habitats), three occur in both forest and savanna, and one occurs primarily in savannas and marginally extends into forests (Table 5.4). The vast majority of bats recorded in this study are typically or predominantly associated with rainforest. One (sub-) species is endemic to Upper Guinea (*Hypsugo* [*crassulus*] *bellieri*), one species is endemic to West Africa (*Epomops buettikoferi*) and two species are near-endemic to West Africa (*Nanonycteris veldkampii*, *Rhinolophus hillorum*). One species might be new to science (*Neoromicia* aff. *grandidieri*).

The smoothed species accumulation curves for Gola and Grebo National Forests are almost identical, indicating similar species-abundance relationships in both assemblages

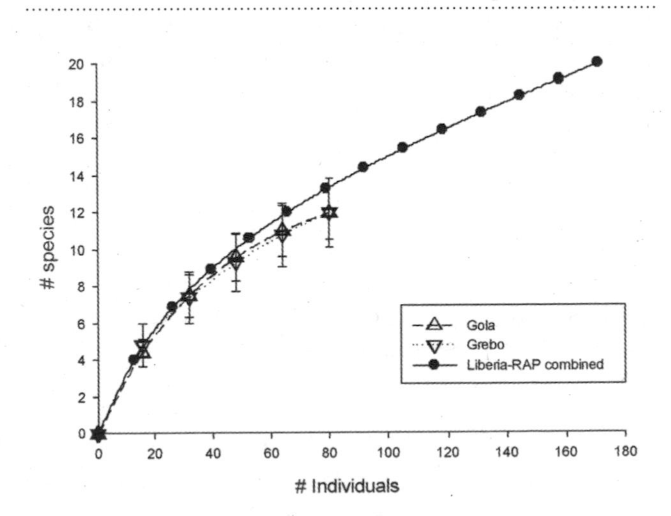

Figure 5.2: Smoothed species accumulation curves for bats sampled during the RAP survey (lines and dots: sample-based rarefaction curves, rescaled by individuals ("Mao Tau"-curves, see Colwell et al. 2004); vertical bars: ± 1 *SD*). Sample size for North Lorma National Forest was too low to calculate a meaningful species accumulation curve.

Table 5.5: Terrestrial small mammal species captured (number) or observed (X) at the three National Forests (see Appendix 9 for specimen details).

Species	Lorma	Gola	Grebo
SORICOMORPHA			
Soricidae			
Crocidura muricauda	1		1
Crocidura obscurior	1		
RODENTIA			
Muridae			
Hylomyscus alleni	1	1	1
Sciuridae			
Epixerus ebii		X	
Funisciurus pyrropus	X	X	
Heliosciurus rufobrachium	X		X
Paraxerus poensis	X	X	
Protoxerus stangeri	X	X	X
Anomaluridae			
Anomalurus cf. *pusillus*			X
Species total	**7**	**5**	**5**

(Figure 5.2). As a result of the low number of bats captured with mist nets in North Lorma National Forest, a species accumulation curve could not be generated for this area. Neither the single-area curves (Gola and Grebo National Forests) nor the overall curve shows a plateau.

Terrestrial small mammals

Three shrews of two species and three murid rodents of one species were collected during this survey (Table 5.5). Due to the limited captures, comparison between the sites cannot be made. In addition to these captures, five species of squirrels (Sciuridae) and one species of scaly-tailed squirrel (Anomaluridae) were observed.

DISCUSSION

The 22 bat species recorded during the RAP survey represent 37% of the 59 species currently known to occur in Liberia, including two species new for the country: *Neoromicia guineensis* and *Neoromicia* aff. *grandidieri*. A third species new for Liberia is based on a previously unpublished museum specimen (*Rhinolophus landeri*; Appendix 10). The species total of seven, 12 and 13 species for each of the surveyed forest areas is within the typical range of RAP surveys. In southeastern Guinea, the survey of four forest reserves resulted in species totals between three and 21 species (Fahr and Ebigbo 2003, 2004; Fahr et al. 2006). The bat survey of three forest sites in southwestern Ghana recorded three, six and 10 species, respectively (Decher and Fahr 2007). The combined species total of all Liberian forest sites (22 species) is comparable to the combined totals from Guinea (21 and 23 species, respectively: Fahr and Ebigbo 2004, Fahr et al. 2006), whereas Ghana has a somewhat lower number (15 species: Decher and Fahr 2007). Extended studies are needed to produce more exhaustive species inventories; at present, only limited comparisons between and qualitative assessments of sites can be made.

Bat capture rates and sampling success critically depend on the specific habitat settings, which explains the sometimes widely differing results. Capture rates of the present RAP survey (0.02–0.16 bats/nh), especially those at North Lorma (0.02 bats/nh), were notably lower than those of previous surveys in West Africa (0.15–1.92 bats/nh: Fahr and Ebigbo 2003, Fahr et al. 2006, Decher and Fahr 2007). Capture success of fruit bats (Pteropodidae) was exceptionally low at North Lorma, and no fruit bats were captured at Gola 1. Capture success of the harp trap was also rather low (0.06 bats/th) compared to that during a RAP survey in southeastern Guinea (1.23 bats/th: Fahr and Ebigbo 2003). However, these results should not be seen as indicating degraded habitat conditions. Capture success in homogeneous, undisturbed rainforest habitat is generally very low compared to situations where habitat mosaic or particular habitat structures can lead to highly increased capture rates (J. Fahr, unpubl. data). This is supported by the results of the present survey: capture success of sites deep in the forest interior (North Lorma and Gola 1) was very low, whereas that of sites on the edge of the forest (Gola 2 and Grebo 2) was much higher. Overall, an exceptionally high capture effort was made during the present survey, which compensated for the low capture success.

At none of the three National Forests can the bat inventories be considered near-complete. The species accumulation curves have not reached an asymptote (Figure 5.2) and further species are thus to be expected. This result parallels those of other RAP surveys, where bat species inventories are far from complete, due to limited sampling effort in space and time (Fahr and Ebigbo 2004, Decher et al. 2005a, Decher and Fahr 2007, Fahr et al. 2006). Fruit bats (Pteropodidae) had the highest species richness and were probably fairly exhaustively sampled. Three species recorded from Liberia, *Scotonycteris ophiodon, Lissonycteris angolensis* and *Eidolon helvum*, are likely to occur in the surveyed National Forests. *Eidolon helvum* is a widely distributed but seasonally migrating species (Thomas 1983) and might have been temporarily absent from the sites during the survey. *Lissonycteris angolensis* has been previously recorded from the Wonegizi and Wologizi Mts. (Koopman et al. 1995; SMNS 39675 – 77) and can be expected to occur in nearby North Lorma National Forest. *Scotonycteris ophiodon* (Endangered) has been recorded in the Mt. Nimba area, as well as near Zwedru and the Dugbe River, southeastern Liberia, and in Taï National Park, Côte d'Ivoire (Fahr in press-e). This species should be the target of additional surveys and can be expected to occur in Grebo National Forest. The inventory of bats in the families Nycteridae, Rhinolophidae, Hipposideridae and Vespertilionidae is far from complete, as revealed when comparing the RAP records with the list of species known to occur in Liberia (Appendix 10). Red List-species, such as *Nycteris major* (Vulnerable = VU), *Rhinolophus guineensis* (VU), *R. ziama* (Endangered = EN) and *Hipposideros marisae* (EN), in particular should be the focus of additional surveys, as many of their records in Liberia are historic and from areas that have since lost their original forest cover. The families Emballonuridae and Molossidae were not recorded during this survey. Bats in these families typically forage in open space above the canopy (Fenton and Griffin 1997). As the highest nets in this study were only 6–8 m above the ground, well below the canopy, the lack of captures of species from these families is due to a sampling bias and does not indicate their absence from the three surveyed forests.

Some of the sampling gaps can be assessed by taking into account species records from locations near the RAP sites. Near Gola National Forest, five additional bat species have been recorded (one from Bomi Wood Concession in Liberia, four from Gola Forest Camp in Sierra Leone; Appendix 11). Noteworthy is the only record of *Myotis tricolor* for West Africa (Koopman 1989). A total of 18 bat species is thus known from the area within 35 km of the RAP site. The Wonegizi Mts. near North Lorma National Forest were intensely surveyed by ornithologist Robert W. Dickerman in March 1990, resulting in a collection of 244 bats (Koopman et al. 1995; AMNH). One of the bats was later described as a species new to science (Fahr et al. 2002). During an earlier survey in April 1972, 25 specimens of seven species were collected in Zigida (USNM). Together, these collections comprise 23–24 species, of which 18–19 species were not recorded in North Lorma National Forest during the present RAP survey (Appendix 11; one record is doubtful). The area within 30 km of the RAP site is thus known to harbor 25–26 bat species, several of which are of conservation concern: *Scotonycteris zenkeri* (Near Threatened = NT), *Rhinolophus hillorum* (VU), *R. guineensis* (VU), *R. ziama* (EN), *Hipposideros marisae* (EN), and *Hipposideros fuliginosus* (NT).

Compared to other West African countries, the list of bat species recorded for Liberia is probably fairly complete (Appendix 10). Nine additional species, known from adjacent localities in neighboring countries, can be expected to occur in Liberia: *Taphozous mauritianus, Nycteris nana, N. gambiensis, Kerivoula cuprosa, Glauconycteris beatrix, G. superba, Chaerephon russatus, C. aloysiisabaudiae,* and *Mops trevori.* In total, Liberia would thus have around 70 bat species.

Threatened and significant bat species

The fruit bat *Scotonycteris zenkeri*, ranked as Near Threatened (IUCN 2006), was recorded from Grebo National Forest during the present study. This species is locally rare to very rare, representing 2.7–3.6% of all fruit bat captures in two extensive studies of local bat assemblages (Fahr in press-f). Although *Scotonycteris zenkeri* is known from several localities throughout Liberia (Kuhn 1965, Verschuren 1977, Wolton et al. 1982, Koopman 1989), most records are

Figure 5.3: Known distribution of *Rhinolophus hillorum* based on records from the RAP survey and Fahr (in press-d). Dark gray: closed forest; light gray: degraded forest and farmland (NOAA/AVHRR-data from 1989; Päivinen et al. 1992).

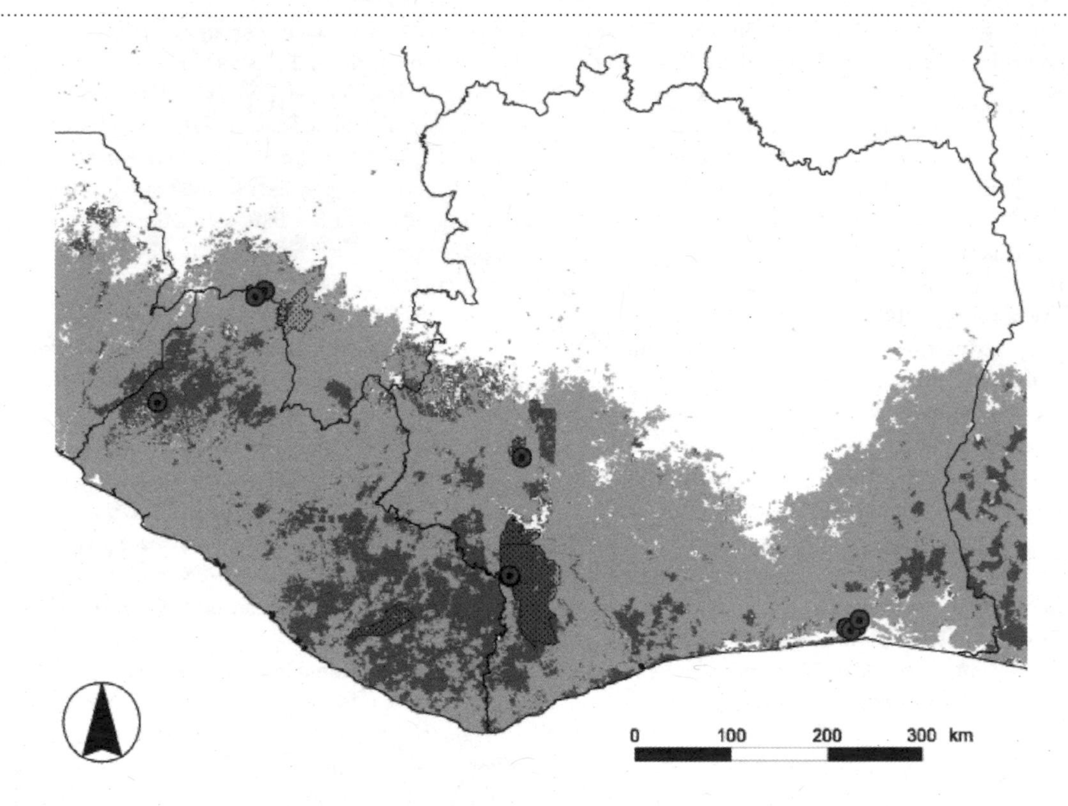

Figure 5.4: Known distribution of *Hypsugo (crassulus) bellieri* based on records from the RAP survey and Fahr (in press-c). Dark gray: closed forest; light gray: degraded forest and farmland (NOAA/AVHRR-data from 1989; Päivinen et al. 1992).

historic and it is likely that this forest-dependent species has disappeared from many of these sites due to degradation and loss of suitable habitat.

Rousettus aegyptiacus, a strictly cave-roosting fruit bat, was previously known in Liberia only from the mountainous regions in the northwest (Wonegizi and Wologizi Mts., Voinjama) and the northeast (Nimba region); the record from Gola National Forest constitutes a southward extension into the Liberian lowland forests.

Rhinolophus hillorum was previously known in Liberia from five localities: River Peblei, south of Grassfield (Verschuren 1977 as *R.* cf. *alcyone*; see below), Tokadeh, Nimba region (Hill 1982 and Wolton et al. 1982 as *R. clivosus*), John's Town, southwest of Voinjama (Koopman 1989 as *R. clivosus hillorum*), and 10.5 km north and 1 km east of Zigida, Wonegizi Mts. (Koopman et al. 1995 as *R. clivosus hillorum* and from "near Ziggida"). There is an additional specimen from the northern foothills of the Wologizi Mts. in the collections of the Stuttgart Museum (SMNS 39671). All of these records originate from the mountainous region near the border with Guinea (Figure 5.3), where *R. hillorum* was recently recorded for the first time (Fahr et al. 2006). The records from Gola National Forest constitute a range extension of approximately 100 km to the southwest. *Rhinolophus hillorum* is near-endemic to West Africa (Figure 5.3) and listed as Vulnerable A4c; B2ab(iii) due to habitat loss within its limited distribution (IUCN 2006). As the species was previously known from only 12 localities and 17 specimens, the two additional localities and three specimens recorded during the present RAP survey are particularly noteworthy.

Rhinolophus alcyone was tentatively recorded from River Peblei by Verschuren (1977), and Happold (1987) listed this species for Liberia without further details, probably based on Verschuren. However, Verschuren's record has been re-identified as *R. hillorum* (Fahr et al. 2006). As no other published records of *R. alcyone* are known from Liberia, the specimens from Grebo National Forest as well as unpublished specimens from Zwedru (SMNS 38526, 38564) constitute the first proof of the species' occurrence in the country. Its presence was to be expected as it was the most frequently captured microbat in nearby Taï National Park (J. Fahr unpubl.) and since it had also been recorded from the Forêt Classée du Cavally (Decher et al. 2005a), both across the border in Côte d'Ivoire.

Hipposideros fuliginosus is a rarely recorded and forest-dependent species. Only 25 localities are known from the forest zone of Upper Guinea (Sierra Leone to Ghana), extreme southeastern Nigeria, Cameroon, and Gabon (Fahr in press-a). Large-sized specimens have been recorded from six localities in the Congolian rainforest zone of D.R. Congo, Central African Republic, and Uganda. *Hipposideros fuliginosus* has been frequently confused with other species of the *H. ruber / caffer* group (Fahr and Ebigbo 2003, Decher and Fahr 2007, Fahr in press-a) and, due to these problems, previous records from Liberia (e.g. Koopman 1989, Koopman et al. 1995) should be critically re-examined. Based on published measurements, however, the record from Voinjama, northwestern Liberia (Koopman 1989), seems indeed referable to *H. fuliginosus*. A series of specimens from Tars Town (USNM 481713, 481715–481718, 481721–481725) was also found to represent this species.

The two large-sized "pipistrelles" captured in Gola National Forest pose considerable taxonomic difficulties as they cannot be referred to any described species known to occur in West Africa. They agree in measurements and characters with four unpublished specimens from Côte d'Ivoire (Coll. Fahr) and a single specimen from extreme southwestern Cameroon (CM 108029: Baké River Bridge). All these specimens in turn agree in measurements and characters with *Neoromicia grandidieri*, described from Zanzibar, East Africa, and currently considered a synonym of *Neoromicia capensis* (Simmons 2005), but resurrected as a distinct species by Thorn et al. (in press). However, the large distributional hiatus between West and East Africa raises the possibility that the West African specimens represent an undescribed taxon. Further morphological and molecular data are needed to answer this question. The record of *Neoromicia* aff. *grandidieri* from Gola National Forest is the first for Liberia.

The taxon *bellieri* is currently recognized as a subspecies of *Hypsugo crassulus* (Heller et al. 1995, Simmons 2005). It has, however, a very restricted distribution within Upper Guinea (Figure 5.4) and probably represents a distinct species (Fahr in press-c). In Liberia, this taxon was recorded from Mt. Klouga (Koopman 1989 as *Pipistrellus eisentrauti bellieri* from "Voinjama") and Balouma, just across the border in Guinea (Fahr et al. 2006). If *bellieri* is ranked as a distinct species, it would qualify as Vulnerable according to the Red List criteria (A4c: IUCN 2006) due to its restricted distribution within the Upper Guinea forest zone (known only from five locations) and extensive habitat degradation and loss.

Terrestrial small mammals

Both shrew species collected during this survey are endemic to the Upper Guinea region. *Crocidura obscurior* is known from Sierra Leone to Côte d'Ivoire and was recently recorded in southwestern Ghana (Decher et al. 2005b). *Crocidura muricauda* is known from the forest zone from Guinea to Ghana. The rodent *Hylomyscus alleni* is known from Guinea (Mt. Nimba) to Bioko and southern Cameroon, but species limits are still unsettled (Musser and Carleton 2005). All three species are closely associated with rainforest, but are widespread and not currently threatened.

Among the recorded squirrels, *Epixerus ebii* is a poorly known species that is rarely seen and whose precise geographic range is not known. As it is likely to be dependent on primary forest, it may be threatened by deforestation; available information, however, does not permit adequate assessment and the species is currently considered Data Deficient (Grubb 2004).

The scaly-tailed squirrel *Anomalurus pusillus* is widely distributed throughout the rainforest zone of Central Africa, from Cameroon, Equatorial Guinea and Gabon in the west to D.R. Congo and Uganda in the east (Schunke 2005). In West Africa, however, this species has been recorded only from Du River, Liberia (Allen and Coolidge 1930), and from an unspecified locality in Côte d'Ivoire (Schunke 2005: 158). The record from Grebo National Forest thus constitutes the only recent record of this species for West Africa. Although *A. pusillus* is ranked as Least Concern (IUCN 2006), the West African population might be threatened by extensive deforestation and its status as an evolutionarily distinct unit should be assessed.

CONCLUSION AND CONSERVATION RECOMMENDATIONS

The rainforest at the three surveyed National Forests appeared to be in good condition. This observation is supported by the fact that most of the recorded bat species are restricted to or mainly found in forest habitat. Given the widespread loss and degradation of forests in Upper Guinea, the Government of Liberia is strongly encouraged to grant the highest possible protection to the surveyed National Forests and to raise both their legal status and management, preferably by creating National Parks that also encompass adjacent forest blocks. Specific recommendations with reference to the small mammal results are given in the following paragraphs.

Although species richness observed during the RAP survey was lowest in North Lorma National Forest, previous records from nearby Wonegizi and Wologizi Mts. include an exceptional number of globally threatened bat species (two Endangered, two Vulnerable, and two Near Threatened; see above). The remaining forest block of North Lorma National Forest forms a critical habitat corridor between these important mountain ranges. It is therefore suggested to create a National Park comprising the entire Wonegizi and Wologizi mountain ranges as well as the lowland forest of North Lorma National Forest. This area should be contiguous to the Biosphere Reserve of the Massif du Ziama in Guinea and would thus constitute one of the most significant protected areas of submontane rainforest in West Africa (for a detailed discussion of the importance of the Massif du Ziama, see Fahr et al. 2006). This mountainous region contains suitable habitat for several cave-roosting bats, many of which have small distribution ranges and are globally threatened.

The bat fauna of Gola National Forest is distinguished by the presence of three species of global conservation relevance: *Rhinolophus hillorum* (VU), *Hypsugo* (*crassulus*) *bellieri* and *Neoromicia* aff. *grandidieri*. Although the latter two taxa have not yet been assessed by the international Red List due to open taxonomic questions, they are restricted to West Africa, known from very few localities, and dependent on undisturbed lowland forest. It is suggested to include Gola National Forest in a larger transboundary area with increased protection, including the remaining lowland forests of Gola Strict Nature Reserve in Sierra Leone and Kpelle National Forest in Liberia.

Grebo National Forest was found to harbor the rare fruit bat *Scotonycteris zenkeri* (NT) as well as the only recent record of the scaly-tailed squirrel *Anomalurus* cf. *pusillus* from West Africa. This area offers opportunities for a transboundary network of protected areas with Côte d'Ivoire, including Taï National Park and the Forêts Classées of Cavally, Goin-Debe and Haute Dodo. Although forest habitat between Grebo National Forest and Taï National Park is broken by a narrow strip of heavy settlement and agriculture, efforts should be undertaken to manage the remaining forest tracts within a larger transboundary framework. It should be noted that the preservation of extensive forest cover is essential to perpetuate the moist air carried by the southwest monsoon further inland. It is highly likely that further loss of forest cover in southeastern Liberia would negatively affect adjacent areas in the north. Only an integrated management of forest reserves in this region of West Africa will safeguard the last remaining patches of unbroken evergreen rainforest.

REFERENCES

Allen, G.M. and H.J. Coolidge. 1930. Mammals of Liberia, *In:* Strong, RP. (ed.).The African Republic of Liberia and the Belgian Congo. Based on the Observations Made and Material Collected During the Harvard African Expedition 1926–1927. Vol. 2. Harvard University Press. Cambridge. Pp. 569–622.

Bakarr, M., B. Bailey, D. Byler, R. Ham., S. Olivieri and M. Omland. (eds.). 2001. From the Forest to the Sea: Biodiversity Connections from Guinea to Togo. Conservation International. Washington, DC.

Bakarr, M., J.F. Oates, J. Fahr, M.P.E. Parren, M.-O. Rödel and R. Demey. 2004. Guinean forests of West Africa, *In:* Mittermeier, R.A., P.R. Gil, M. Hoffman, J. Pilgrim, T. Brooks, C.G. Mittermeier, J. Lamoreux and G.A.B. Da Fonseca (eds.). Hotspots Revisited: Earth's Biologically Richest and Most Endangered Terrestrial Ecoregions. CEMEX / Agrupación Sierra Madre. Mexico City. Pp. 123–130.

Bayol, N. and J.-F. Chevalier. 2004. Current State of the Forest Cover in Liberia: Forest Information Critical to Decision Making. Final report to the World Bank. Forêt Ressources Management. Mauguio, France.

Beaucournu, J.-C. and J. Fahr. 2003. Notes sur les Ischnopsyllinae du Continent Africain. IV: Quelques *Lagaropsylla* JORDAN & ROTHSCHILD 1921 de Côte d'Ivoire; description de *L. senckenbergiana* n. sp. (Insecta: Siphonaptera: Ischnopsyllidae). Senckenbergiana biol. 82: 157–162.

Bergmans, W. 1988. Taxonomy and biogeography of African fruit bats (Mammalia, Megachiroptera). 1. General introduction; material and methods; results: The genus *Epomophorus* BENNET, 1836. Beaufortia 38(5): 75–146.

Coe, M. 1975. Mammalian ecological studies on Mount Nimba, Liberia. Mammalia 39: 523–588.

Colwell, R.K. 2005. EstimateS: Statistical Estimation of Species Richness and Shared Species from Samples. Version 7.5. Application and User's guide. Website: purl.oclc.org/estimates.

Colwell, R.K., C.X. Mao and J. Chang. 2004. Interpolating, extrapolating, and comparing incidence-based species accumulation curves. Ecology 85: 2717–2727.

Csorba, G., P. Ujhelyi and N. Thomas. 2003. Horseshoe Bats of the World (Chiroptera: Rhinolophidae). Alana Books. Bishop's Castle, Shropshire, UK.

Decher, J. and J. Fahr. 2007 (in press). A conservation assessment of bats (Chiroptera) of Draw River, Boi-Tano, and Krokosua Hills Forest Reserves in the Western Region of Ghana. Myotis 43.

Decher, J., B. Kadjo, M. Abedi-Lartey, E.O. Tounkara and S. Kante. 2005a. A rapid survey of small mammals (shrews, rodents, and bats) from the Haute Dodo and Cavally Forests, Côte d'Ivoire, *In:* Alonso, L.E., F. Lauginie, and G. Rondeau (eds). A Rapid Biological Assessment of Two Classified Forests in South-Western Côte d'Ivoire. RAP Bulletin of Biological Assessment 34. Conservation International. Washington, DC. Pp.101–109.

Decher, J., J. Oppong and J. Fahr. 2005b. Rapid assessment of small mammals at Draw River, Boi-Tano, and Krokosua Hills, *In:* McCullough, J., J. Decher, and D. Guba Kpelle (eds.). A Biological Assessment of the Terrestrial Ecosystems of the Draw River, Boi-Tano, Tano Nimiri and Krokosua Hills Forest Reserves, Southwestern Ghana. RAP Bulletin of Biological Assessment 36. Conservation International. Washington, DC. Pp. 57–66, 151–152.

DIVA-GIS, software version 5.2.0.2. Website: www.diva-gis.org.

Fahr, J. (in press-a). *Hipposideros fuliginosus. In:* Happold, D.C.D., J. Kingdon, and T. Butynski (eds.). The Mammals of Africa. Vol. 3. Elsevier Science and Academic Press. Amsterdam and London.

Fahr, J. (in press-b). *Kerivoula smithii. In:* Happold, D.C.D., J. Kingdon, and T. Butynski (eds.). The Mammals of Africa. Vol. 3. Elsevier Science and Academic Press. Amsterdam and London.

Fahr, J. (in press-c). *Pipistrellus crassulus. In:* Happold, D.C.D., J. Kingdon, and T. Butynski (eds.). The Mammals of Africa. Vol. 3. Elsevier Science and Academic Press. Amsterdam and London.

Fahr, J. (in press-d). *Rhinolophus hillorum. In:* Happold, D.C.D., J. Kingdon, and T. Butynski (eds.). The Mammals of Africa. Vol. 3. Elsevier Science and Academic Press. Amsterdam and London.

Fahr, J. (in press-e). *Scotonycteris ophiodon. In:* Happold, D.C.D., J. Kingdon, and T. Butynski (eds.). The Mammals of Africa. Vol. 3. Elsevier Science and Academic Press. Amsterdam and London.

Fahr, J. (in press-f). *Scotonycteris zenkeri. In:* Happold, D.C.D., J. Kingdon, and T. Butynski (eds.). The Mammals of Africa. Vol. 3. Elsevier Science and Academic Press. Amsterdam and London.

Fahr, J., B.A. Djossa, and H. Vierhaus. 2006. Rapid assessment of bats (Chiroptera) in Déré, Diécké and Mt. Béro classified forests, southeastern Guinea; including a review of the distribution of bats in Guinée Forestière. *In:* Wright, H.E., J. McCullough, L.E. Alonso, and M.S. Diallo (eds.). A Rapid Biological Assessment of Three Classified Forests in Southeastern Guinea. RAP Bulletin of Biological Assessment 40. Conservation International. Washington, DC. Pp. 168–180, 245–247.

Fahr, J. and N.M. Ebigbo. 2003. A conservation assessment of the bats of the Simandou Range, Guinea, with the first record of *Myotis welwitschii* (Gray, 1866) from West Africa. Acta Chiropterologica 5: 125–141.

Fahr, J. and N.M. Ebigbo. 2004. Rapid survey of bats (Chiroptera) in the Forêt Classée du Pic de Fon, Guinea. *In:* McCullough, J. (ed.). A Rapid Biological Assessment of the Forêt Classée du Pic de Fon, Simandou Range, South-eastern Republic of Guinea. RAP Bulletin of Biological Assessment 35. Conservation International. Washington, DC. Pp. 69–77.

Fahr, J., H. Vierhaus, R. Hutterer and D. Kock. 2002. A revision of the *Rhinolophus maclaudi* species group with the description of a new species from West Africa (Chiroptera: Rhinolophidae). Myotis 40: 95–126.

FAO. 2006. Global Forest Resources Assessment 2005. Progress Towards Sustainable Forest Management. FAO Forestry Paper N° 147.

Fenton, M.B. and D.R. Griffin. 1997. High-altitude pursuit of insects by echolocating bats. *J. Mamm.* 78: 247–250.

Grubb, P. 2004. *Epixerus ebii. In:* 2006 IUCN Red List of Threatened Species. Website: www.iucnredlist.org.

Grubb, P., T.S. Jones, A.G. Davies, E. Edberg, E.D. Starin and J.E. Hill. 1999. Mammals of Ghana, Sierra Leone, and The Gambia. The Trendine Press. Zennor, St. Ives.

Happold, D.C.D. 1987. The Mammals of Nigeria. Clarendon. Oxford.

Heller, K.-G., M. Volleth and D. Kock. 1995 [for 1994]. Notes on some vespertilionid bats from the Kivu region, Central Africa (Mammalia: Chiroptera). Senckenbergiana biol. 74: 1–8.

Hijmans, R.J., S.E. Cameron, J.L. Parra, P.G. Jones and A. Jarvis. 2005. Very high resolution interpolated climate surfaces for global land areas. International Journal of Climatology 25: 1965–1978.

Hill, J.E. 1982. Records of bats from Mount Nimba, Liberia. Mammalia 46: 116–120.

Howard, P.C., T.R.B. Davenport, F.W. Kigenyi, P. Viskanic, M.C. Baltzer, C.J. Dickinson, J. Lwanga, R.A. Matthews and E. Mupada. 2000. Protected area planning in the tropics: Uganda's national system of forest nature reserves. Conserv. Biol. 14: 858–875.

International Tropical Timber Organization (ITTO). 2006. Status of Tropical Forest Management 2005. ITTO Technical Series N° 24.

IUCN. 2006. 2006 IUCN Red List of Threatened Species. Website: www.iucnredlist.org. Downloaded May 2006.

Jentink, F.A. 1888 [for 1887]. Zoological researches in Liberia. A list of mammals, collected by J. Büttikofer, C. F. Sala and F. X. Stampfli, with biological observations. Notes Leyden Mus. 10: 1–58.

Kock, D. 2001. Identity of the African *Vespertilio hesperida* TEMMINCK 1840 (Mammalia, Chiroptera, Vespertilionidae). Senckenbergiana biol. 81: 277–283.

Koopman, K.F. 1989. Systematic notes on Liberian bats. Am. Mus. Novitates 2946: 1–11.

Koopman, K.F., C.P. Kofron and A. Chapman. 1995. The bats of Liberia: Systematics, ecology, and distribution. Am. Mus. Novitates 3148: 1–24.

Kuhn, H.-J. 1962. Zur Kenntnis der Microchiroptera Liberias. Zool. Anz. 168: 179–187.

Kuhn, H.-J. 1965. A provisional checklist of the mammals of Liberia (With notes on the status and distribution of some species.). Senckenbergiana biol. 46: 321–340.

Küper, W., J.H. Sommer, J.C. Lovett, J. Mutke, H.P. Linder, H.J. Beetje, R.S.A.R. Van Rompaey, C. Chatelain, M. Sosef and W. Barthlott. 2004. Africa's hotspots of biodiversity redefined. Ann. Missouri Bot. Gard. 91: 525–535.

Martin, R.E., R.H. Pine and A.F. DeBlase. 2001. A Manual of Mammalogy with Keys to Families of the World. 3rd edn. McGraw-Hill Publishing. Boston.

Miller jr., G. S. 1900. A collection of small mammals from Mount Coffee, Liberia. Proc. Wash. Acad. Sci. 2: 631–649.

Musser, G.G. and M.D. Carleton. 2005. Superfamily Muroidea, *In:* Wilson, D.E. and D.M. Reeder (eds.). Mammal Species of the World: A Taxonomic and Geographic Reference. Vol. 2. Johns Hopkins University Press. Baltimore. Pp. 894–1531.

Myers, N., R.A. Mittermeier, C.G. Mittermeier, G.A.B. da Fonseca and J. Kent. 2000. Biodiversity hotspots for conservation priorities. Nature 403: 853–858.

Päivinen, R., J. Pitkänen and R.G. Witt. 1992. Mapping closed tropical forest cover in West Africa using NOAA/AVHRR-LAC data. Silva Carelica 21: 25–52.

Robbins, C.B., F. De Vree and V. Van Cakenberghe. 1985. A systematic revision of the African bat genus *Scotophilus* (Vespertilionidae). Ann. Mus. roy. Afr. Centr. (Sci. zool.) 246: 53–84.

Schunke, A.C. 2005. Systematics and Biogeography of the African Scaly-tailed Squirrels (Mammalia: Rodentia: Anomaluridae). Unpublished Ph.D. thesis. Bonn: Rheinische Friedrich-Wilhelms-Universität.

Simmons, N.B. 2005. Order Chiroptera, *In:* Wilson, D.E. and D.M. Reeder (eds.). Mammal Species of the World: A Taxonomic and Geographic Reference. Vol. 1. Johns Hopkins University Press. Baltimore. Pp. 312–529.

Thomas, D.W. 1983. The annual migrations of three species of West African fruit bats (Chiroptera: Pteropodidae). Can. J. Zool. 61: 2266–2272.

Thorn, E., D. Kock and J. Cuisin, J. (in press). Status of the African bats *Vesperugo grandidieri* DOBSON 1876 and *Vesperugo flavescens* SEABRA 1900 and description of a new genus *Afropipistrellus* (Chiroptera, Vespertilionidae). Mammalia.

Verschuren, J. 1977 [for 1976]. Les cheiroptères du Mont Nimba (Liberia). Mammalia 40: 615–632.

Voss, R.S. and L.H. Emmons. 1996. Mammalian diversity in Neotropical lowland rainforests: A preliminary assessment. Bull. Am. Mus. Nat. Hist. 230: 1–115.

Wilson, D.E. and D.M. Reeder (eds.). 2005. Mammal Species of the World: A Taxonomic and Geographic Reference. 3rd edn. Johns Hopkins University Press. Baltimore.

Wolton, R.J., P.A. Arak, H.C.J. Godfray and R.P. Wilson. 1982. Ecological and behavioural studies of the Megachiroptera at Mount Nimba, Liberia, with notes on Microchiroptera. Mammalia 46: 419–448.

Chapter 6

Rapid survey of large mammals of North Lorma, Gola and Grebo National Forests

Abdulai Barrie, Sormongar Zwuen, Aaron N. Kota, Sr., Miaway Luo and Roger Luke

SUMMARY

A Rapid Assessment Program survey was conducted from 16 November to 12 December 2005, to record the presence of large mammals, including primates, in three Liberian National Forests. Tracks, sound and visual observations and camera phototraps were used in the survey. During the 15 days of field work 29 mammal species were recorded: 21 in North Lorma National Forest, 14 in Gola National Forest and 28 in Grebo National Forest. Nine were primate species, including one prosimian (Demidoff's Galago *Galagoides demidoff*), seven anthropoid monkeys (Sooty Mangabey *Cercocebus atys*, Campbell's Monkey *Cercopithecus campbelli*, Lesser Spot-nosed Monkey *C. petaurista*, Diana Monkey *C. diana*, Western Red Colobus *Piliocolobus badius*, Western Pied Colobus *Colobus polykomos* and Olive Colobus *Procolobus verus*) and one hominoid ape (West African Chimpanzee *Pan troglodytes verus*). Three of these primate species are listed on the IUCN Red List as Endangered (*Pan troglodytes verus*, *Piliocolobus badius* and *Cercopithecus diana*) or Near Threatened (*Colobus polykomos*, *Procolobus verus* and *Cercocebus atys*). Other large mammal species of conservation concern that were recorded include Forest Elephant *Loxodonta africana cyclotis*, Pygmy Hippopotamus *Hexaprotodon liberiensis*, Leopard *Panthera pardus*, Bongo *Tragelaphus euryceros*, Bay Duiker *Cephalophus* dorsalis, Jentink's Duiker *C. jentinki*, Maxwell's Duiker *C. maxwelli*, Black Duiker *C. niger*, Ogilby's Duiker *C. ogilbyi* and Yellow-backed Duiker *C. silvicultor*. All of the forests were active timber concessions before the war in 1989. Artisanal mining was observed in Gola National Forest and prospecting for large-scale mining is occurring. Although hunting in National Forests is prohibited in Liberia, evidence of poaching was found in all three forests. Despite human pressures, North Lorma, Gola and Grebo National Forests still contain a wealth of large mammal diversity and should be protected.

INTRODUCTION

Primates and other large mammals are indicators of the biodiversity and state of a site and represent an important part of tropical ecosystems (Davies and Hoffmann 2002). North Lorma, Gola, and Grebo National Forests are part of the Upper Guinea hotspot, which includes forests from eastern Sierra Leone to eastern Togo and is considered one of the world's 34 priority conservation areas because of its high degree of biodiversity and endemism (Mittermeier et al. 2004).

Primate densities are high in some forests in the region (Whitesides et al. 1988, Struhsaker and Bakarr 1999, Kormos and Boasch 2003). However, large mammals are highly threatened as a result of the dramatic rate of deforestation which has caused the loss of nearly 75% of the original forest cover (Bakarr et al. 2001). Habitat loss and high hunting pressure account for the loss or reduction of species in the West African forests (Oates 1986, Lee et al. 1988, Bakarr et al. 2001, Kingdon 1997). Many large mammals, including primates, have declined drastically and some forms have been completely extirpated in certain countries (e.g., the recent extinction of Miss Waldron's Red Colobus *Piliocolobus badius waldroni* in Ghana: Oates et al. 2000). The West African Chimpanzee is believed to be extinct in four West African countries (The Gambia, Burkina Faso, Togo and Benin) and Liberia is one of the few countries, along with Guinea, Sierra Leone, Côte d'Ivoire and Mali, with viable populations (Kormos and Boesch 2003).

The Forest Elephant *Loxodonta africana cyclotis* is found in very small and relict populations in both savanna and forest, and is a species of particular concern due to degradation and fragmentation of suitable habitats (Barnes 1999). Other important large mammals of conservation concern include Leopard *Panthera pardus*, Bongo *Tragelaphus euryceros*, Pygmy Hippopotamus *Hexaprotodon liberiensis*, Zebra Duiker *Cephalophus zebra* and Jentink's Duiker *C. jentinki.*

At a time when deforestation and bushmeat hunting are increasing across Upper Guinea, survey information is particularly important to assess species diversity and density, and monitor long-term effects of habitat changes. As part of a strategy to protect biodiversity, Conservation International undertook a rapid survey of North Lorma, Gola and Grebo National Forests, to support Liberia's Forest Reassessment Programme and to provide appropriate data to set priorities for biodiversity conservation in Liberia and the Upper Guinea region.

METHODS

Surveys were conducted at the end of the rainy season in North Lorma National Forest (08°01'53.6"N, 09°44' 08.6"W) from 20 to 24 November 2005, Gola National Forest (07°27'09.9"N, 10°41'33.2"W) from 28 November 28 to 3 December 2005 and Grebo National Forest (05°24'10.4"N, 07°43'56.2"W) from 6 to 11 December 2005. The three sites were approximately 400m a.s.l. National forests in Liberia are set aside for timber production and the three forests had all been logged before the civil war in 1989. Artisanal mining was observed in Gola National Forest during the survey and two mining companies were prospecting in order to start full-scale mining operations in anticipation of a lifting of the sanctions imposed by the United Nations on Liberia. Forest streams and rivers were flowing and it rained regularly during our survey, most frequently in North Lorma National Forest.

Active and passive methods were used to document the presence of large nonvolant mammals and to count the numbers of individuals when possible. Active methods included direct observations of animals and sounds and indirect information such as dens/nests, dung, tracks, feeding sites and rooting. Observations were made during daily excursions from base camp during the day and at night (when a spotlight was used). Some observations were made opportunistically by our colleagues, but as these may have been repeats, we used this information only to document species presence, rather than adding these records to our counts of individuals.

The passive method included the use of CamTrakker cameratraps, triggered by heat-in-motion (CamTrakker Atlanta, Georgia). Thirteen were used at North Lorma National Forest and seven in both Gola and Grebo National Forests. Each CamTrakker was equipped with a Samsung Vega 77i 35mm camera set on autofocus and loaded with Fujicolor Superia 200 print film. Time between sensor reception and the taking of a photograph was 0.6 seconds. Cameras were set to operate continuously (control switch 1 on) and to wait at least 20 seconds between photographs (control switches 6 and 8 on). They were placed at sites suspected of being frequented by various mammalian species, such as dens, trails, and feeding stations, particularly fruiting trees. Cameras were located approximately 500 m apart and at least 1000 m from base camp. We used this method to calculate observation rates for each site. Instead of the observer making observations along a standard transect, "observations" moved along routes in front of fixed cameras (observers). For shy mammals under severe hunting pressure camera-trapping might be more effective than walking transects, especially when observers have different levels of expertise (Sanderson and Trolle 2005, Karanth and Nicholas 1998).

RESULTS

We observed, heard or photographed a total of 29 large mammal species: 21 in North Lorma National Forest, 14 in Gola National Forest, and 28 in Grebo National Forest (Appendix 12). Among these were four Endangered mammal species (West African Chimpanzee *Pan troglodytes verus*, Western Red Colobus *Piliocolobus badius*, Diana Monkey *Cercopithecus diana* and Pygmy Hippopotamus *Hexaprotodon liberiensis*); two Vulnerable species (Forest Elephant *Loxodonta africana cyclotis* and Jentink's Duiker *Cephalophus jentinki*); one Lower Risk/Conservation Dependant species (African Buffalo *Syncerus caffer*); and nine Lower Risk/Near Threatened species (Western Pied Colobus *Colobus polykomos*, Olive Colobus *Procolobus verus*, Sooty Mangabey *Cercocebus atys*, Bay Duiker *Cephalophus dorsalis*, Maxwell's Duiker *C. maxwelli*, Black Duiker *C. niger*, Ogilby's Duiker *C. ogilbyi*, Yellow-backed Duiker *C. silvicultor* and Bongo *Tragelaphus euryceros*) (IUCN 2006). The cameratraps obtained two photographs in North Lorma National Forest, one in Gola National Forest, and three in Grebo National Forest. The six photographs included three of Maxwell's Duiker *Cephalophus maxwelli* and one each of Black Duiker *C. niger*, Yellow-backed Duiker *C. silvicultor* and Jentink's Duiker *C. jentinki.* One photograph of a ground-dwelling bird (White-crested Tiger Heron *Tigriornis leucolopha*) was also taken.

With the observation of tracks and dung, and the use of cameratraps, we recorded the presence of Pygmy Hippopotamus, Forest Elephant, and Leopard in North Lorma and Grebo National Forests. In these two forests we observed primates every day, whereas in Gola National Forest we did so on only four occasions. We recorded eight primate species in North Lorma National Forest, three in Gola National Forest, and nine in Grebo National Forest. In North Lorma National Forest, one old and seven rotten tree nests confirmed the continued presence of West African Chimpanzees, and in Grebo National Forest we heard their vocalizations (once) and drumming (daily), but direct sightings were not made. In Grebo National Forest we also found West African Chimpanzee nut-cracking sites. The fruits cracked were mainly those of *Parinari excelsa*. During an encounter with a large group of Western Red Colobus and Diana and Campbell's Monkeys, West African Chimpanzee calls sent the monkeys looking for cover and remaining silent for about 30 minutes. No evidence of West African Chimpanzees was found in Gola National Forest but hunters reported that they still occurred in some parts of the forest.

African Buffalo *Syncerus caffer*, Red River Hog *Potamochoerus porcus* and Olive Colobus *Procolobus verus* were documented only from North Lorma National Forest. This is probably due to the short duration of the survey and not to fundamental differences in mammalian faunas in the study areas.

Although Royal Antelope *Neotragus pygmaeus*, Zebra Duiker *Cephalophus zebra* and Aardvark *Orycteropus afer* were not recorded, local hunters reported that these species still occurred in North Lorma and Grebo National Forests. Hunters also reported the latter two at Gola National Forest, as well as other species we did not observe there, such as West African Chimpanzee, Western Red Colobus, Western Pied Colobus, and Pangolin.

DISCUSSION

The diversity and density of large mammals recorded during the present survey is high compared to results from other RAP surveys in Guinea, Côte d'Ivoire, and Ghana (Struhsaker and Bakarr 1999, Barrie and Kante 2004, Herbinger and Tounkara 2004, Sanderson and Trolle 2005, Barrie and Aalangdong 2005). This offers high potential for the conservation of primates and other large mammals in the Upper Guinean forest region. Although our survey lasted less than a month, we noted many species of conservation concern in reasonable numbers (Appendix 12), among which were six primates (*Cercocebus atys, Cercopithecus diana, Piliocolobus badius, Colobus polykomos, Procolobus verus* and *Pan troglodytes verus*) and six other large mammal species (*Hexaprotodon liberiensis, Cephalophus jentinki, C. niger, C. silvicultor, C. dorsalis, Syncerus caffer, Tragelaphus euryceros* and *Panthera pardus*). This offers hope for the future of large mammal species in Liberia.

However, the protected areas on which these species depend may become too small, fragmented and overexploited for their long-term survival, as these forests are threatened by human actions including commercial logging, mining, agricultural activities and bushmeat trade. Current distribution patterns of most large mammals and observed human activities in the areas under investigation in Liberia reflect these increasing pressures.

Logging is locally intense and destructive in many countries in West Africa and has been cited as the primary cause of habitat destruction in Sierra Leone (Bakarr et al. 2001). Primary forests outside protected areas are targeted for timber extraction and secondary forests are being encroached upon. Before the war, Liberia was a major source of timber and this has caused, and will continue to cause, forest fragmentation and the subsequent loss of large mammals. In particular, the Gola National Forest was being surveyed for commercial logging and mining. The Liberian civil conflict also negatively impacted upon large mammals as most of the forests were abandoned by government authorities and plundered by rebels engaged in illegal mining and logging.

Secondary impacts of resource extraction such as roads and trails are equally destructive to the forest. Logging roads create easy access for hunters (and others) into areas that were otherwise not penetrable (Sayer et al. 1992, Wilkie et al. 1992, Oates 1999). As humans move deeper into the forests, diseases to wildlife increase, especially to primates that have had no previous contact with humans (Chapman et al. 2006). Workers often support themselves and their families on bushmeat, consume trees for fuel wood and clear areas to plant crops. In addition, the increase in human population is accelerating the conversion of remaining forest habitats into human-dominated settlements and agricultural landscapes. Local communities cause additional habitat degradation by establishing farms and hunting within the boundaries of forest reserves.

During the survey we used extensive networks of trails created by heavy machinery and poachers. The roads and trails fragment the forest reducing the area for wildlife. Collateral damage from logged trees was extensive and many untargeted trees had been damaged. We saw evidence of trees that were cut and abandoned.

As the human population grows, bushmeat markets develop. Large mammals, including primates, are extremely rare in much of West Africa as a result of unregulated exploitation, habitat loss and the increasing demand for bushmeat (Davies 1987, Grubb et al. 1998, McGraw 1998, Davies and Hoffmann 2002). Populations of forest-dependent animals have been reduced to such low levels that a number of them can no longer be considered viable. Large mammals, prime targets for the bushmeat trade, are usually the first to be eliminated from forest areas. As in most other countries in West and Central Africa, people in Liberia have always hunted and relied on bushmeat to provide them with protein. Species most preferred by hunters include antelopes, forest pigs, and primates, while smaller species like the Cane Rat *Thryonomys swinderianus* and Giant Rat *Cricetomys* spp. are taken opportunistically (Eves and Bakarr 2001). The bushmeat trade is a lucrative business in Liberia, as in other parts of Africa (Oates 1986, Barrie and Kante 2004, Sanderson and Trolle 2005, Barrie and Aalandong 2005), and the amount of bushmeat coming out of forest reserves continues to increase, despite laws banning hunting. The lack of law enforcement is often a major problem. The apparent extinction of Miss Waldron's Red Colobus *Piliocolobus badius waldroni* has been attributed to hunting and the demand for bushmeat (Oates et al. 2000). West African Chimpanzees are the most threatened of the three subspecies mainly due to habitat loss, high hunting pressure and the pet trade (Kormos and Boesch 2003). Around the forest reserves surveyed during the RAP, hunting was found to be a major source of meat. We found two Sooty Mangabeys and a Marsh Mongoose being kept by local villagers and through conversations it was apparent that these animals would be eaten.

Despite the various threats, Gola, Lorma and Grebo National Forests contain an important representation of the mammalian diversity of the region and Liberia and thus the inclusion of these forests into a protected area system has great potential for the conservation of these species in the Upper Guinea region.

CONSERVATION RECOMMENDATIONS

Raise both North Lorma and Grebo National Forests to National Park status. Eight primate species were recorded in North Lorma National Forest and nine primate species in Grebo NF. In comparison, Sapo National Park, which is the "core of an immense forest block that has not been disturbed or fragmented to the same extent as most of the Upper Guinea Forest Ecosystem, and as such it offers fantastic conservation opportunities" (Waitkuwait 2001), is known to support nine species of primates including West African Chimpanzees (Waitkuwait 2003). Of the mammal species noted in both forests, over half are of conservation concern including 63% of the primate species in North Lorma National Forest and 67% in Grebo National Forest. In Grebo National Forest primate populations of Western Red Colobus, Western Pied Colobus and Diana Monkeys were numerous and were seen daily in large groups. In addition, the presence of Pygmy Hippopotamus, Jentink's Duiker, Forest Elephant, Leopard and Bongo make this forest very important for large mammal conservation.

Create a biological corridor to connect the Gola National Forest in Liberia and the Gola Forest in Sierra Leone. Both Sierra Leone and Liberia have had very brutal civil conflicts and both countries can now work towards making a joint effort aimed at protecting biodiversity through the creation of a "Peace Park". This would not only enhance conservation efforts but could also help to maintain and foster peace and stability between the two countries.

Further surveys are strongly recommended for this site as the areas investigated were relatively degraded due to mining and poaching. Despite the degradation, five species of conservation concern were recorded with hunters indicating that additional species, such as West African Chimpanzee, were present in the forest.

With the formation of a transboundary park, Liberia and Sierra Leone could undertake a joint monitoring program of migrant and threatened mammal species such as West African Chimpanzee, Forest Elephant, Pygmy Hippopotamus, Bongo, Leopard, etc.

Halt all human activities that exploit and damage the forest and wildlife (e.g. logging, mining, hunting). Logging and mining interests are currently sizing up the forests for resource extraction. Hunters are using networks of old logging roads and poaching trails to kill wildlife for food and the bushmeat trade.

True protection for Liberia's forest and large mammal populations is not possible without the support of the surrounding communities. Therefore, the following is also recommended for North Lorma, Gola and Grebo National Forests:

Establish a conservation education and awareness program involving the local communities so that people know and understand the importance of the forest and biodiversity conservation.

Establish community forest monitors and wildlife guards and train them in patrolling techniques. Regular monitoring of the forest by monitors and guards is necessary to deal with illegal hunting and trade in bushmeat which is occurring.

Conduct further survey work during different seasons to get a complete picture of the diversity and abundance of the large mammal species in the three reserves.

Monitor species of conservation concern: West African Chimpanzee *Pan troglodytes verus*, Western Red Colobus *Piliocolobus badius*, Diana Monkey *Cercopithecus diana*, Pygmy Hippopotamus *Hexaprotodon liberiensis*, Forest Elephant *Loxodonta africana cyclotis*, Jentink's Duiker *Cephalophus jentinki*, African Buffalo *Syncerus caffer*, Western Pied Colobus *Colobus polykomos*, Olive Colobus *Procolobus verus*, Sooty Mangabey *Cercocebus atys*, Bay Duiker *Cephalophus dorsalis*, Maxwell's Duiker *Cephalophus maxwelli*, Black Duiker *Cephalophus niger*, Ogilby's Duiker *Cephalophus ogilbyi*, Yellow-backed Duiker *Cephalophus silvicultor*, Bongo *Tragelaphus euryceros* and Leopard *Panthera pardus*. This could be done in collaboration with Liberian universities, NGOs and other research institutions. Field or research stations can also act as deterrents to hunters and other illegal activities.

REFERENCES

Bakarr, M.I., B. Baily, D. Byler, R. Ham, S. Olivieri and M. Omland (eds.). 2001. From the Forest to the Sea: Biodiversity Connections from Guinea to Togo. Conservation International. Washington, DC.

Bakarr, M.I., G.A.B. da Fonseca, R.A. Mittermeier, A.B. Rylands and K. Walker Painemilla (eds.). 2001. Hunting and Bushmeat Utilization in the African Rain Forest: Perspectives toward a Blueprint for Conservation Action. Advances in Applied Biodiversity Science 2. Center for Applied Biodiversity Science, Conservation International. Washington, DC.

Barnes, R.F.W. 1999. Is there a future for elephants in West Africa? Mammal Review 29: 175–199.

Barrie, A. 2002. Post conflict conservation status of large mammals in the Western Area Forest Reserve (WAFR), Sierra Leone. Unpublished M. Sc. thesis. Freetown, Sierra Leone: Njala University College.

Barrie, A. and O.I. Aalangdong. 2005. Rapid assessment of large mammals at Draw River, Boi-Tano and Krokosua Hills. *In*: McCullough, J., J. Decher, and D. Guba Kpelle (eds.). A Biological Assessment of the Terrestrial Ecosystems of the Draw River, Boi-Tano, Tano Nimiri and Krokosua Hills Forest Reserves, Southwestern Ghana. RAP Bulletin of Biological Assessment 36. Conservation International. Washington, DC. Pp. 67–72, 153.

Barrie, A. and S. Kante. 2004. A rapid survey of the large mammals of the Forêt Classée du Pic de Fon, Guinea. *In:* McCullough, J. (ed.). A Rapid Biological Assessment of the Forêt Classée du Pic de Fon, Simandou Range, South-eastern Republic of Guinea. RAP Bulletin of Biological Assessment 35. Conservation International. Washington, DC. Pp. 84–90.

Chapman, C.A., M.J. Lawes and H.A.C. Eeley. 2006. What hope for African primate diversity? African Journal of Ecology 44:116–133.

Davies, A.G. 1987. Conservation of primates in the Gola Forest reserves, Sierra Leone. Primate Conservation 8: 151–153.

Davies, G. and M. Hoffmann (eds.). 2002. African Forest Biodiversity. A Field Survey Manual for Vertebrates. Earthwatch Europe. UK.

Eves, H.E. and M.I. Bakarr. 2001. Impacts of bushmeat hunting on wildlife populations in West Africa's Upper Guinea Forest Ecosystem. *In:* Bakarr, M.I., G.A.B. da Fonseca, R. Mittermeier, A.B. Rylands and K.W. Painemilla (eds.). Hunting and Bushmeat Utilization in the African Rain Forest: Perspectives toward a Blueprint for Conservation Action. Advances in Applied Biodiversity Science 2. Conservation International, Washington, DC. Pp. 39–57.

Grubb, P., T.S. Jones, A.G. Davies, E. Edberg, E.D. Starin and J.E. Hill. 1998. Mammals of Ghana, Sierra Leone and The Gambia. The Tendrine Press. Zennor, St Ives.

Herbinger, I. and E.O. Tounkara. 2004. A rapid survey of primates in the Forêt Classée du Pic de Fon, Guinea. *In*: McCullough, J. (ed.). A Rapid Biological Assessment of the Forêt Classée du Pic de Fon, Simandou Range, South-eastern Republic of Guinea. RAP Bulletin of Biological Assessment 35. Conservation International. Washington, DC. Pp. 91–99

IUCN 2006. 2006 IUCN Red List of Threatened Species. Web site: www.iucnredlist.org. Downloaded on 23 January 2007.

Karanth, K.S. and J.D. Nicholas. 1998. Estimation of tiger densities in India using photographic captures and recaptures. Ecology 79: 2852–2862.

Kingdon, J. 1997. The Kingdon Field Guide to African Mammals. Academic Press. San Diego.

Kormos, R. and C. Boesch. 2003. Regional Action Plan for the Conservation of Chimpanzees in West Africa. IUCN/SSC Action Plan. Conservation International. Washington, DC.

Lee, P.C., J. Thornback and E.L. Bennett. 1988. Threatened Primates of Africa. The IUCN Red Data Book. IUCN. Gland, Switzerland and Cambridge, UK.

McGraw, W.S. 1998. Three monkeys nearing extinction in the forest reserves of eastern Côte d'Ivoire. Oryx 32: 233–236.

Mittermeier, R.A., P. Robles Gil, M. Hoffmann, J. Pilgrom, T. Brooks, C.G. Mittermeier, J. Lamoreux and G.A.B. da Fonseca (eds.). 2004. Hotspots Revisited. Earth's Biologically Richest and Most Endangered Terrestrial Ecoregions. CEMEX/Agrupación Sierra Madre, Mexico City.

Oates, J.F. 1986. Action Plan for African Primate Conservation 1986–1990. IUCN/SSC Primate Specialist Group. New York.

Oates, J.F. 1999. Myth and Reality in the Rainforest: How Conservation Strategies are Failing in Africa. University of California Press. Berkeley.

Oates, J.F., M. Abedi-Lartey, S. McGraw, T.T. Struhsacker and G.H. Whitesides. 2000. Extinction of a West African red colobus monkey. Conservation Biology 14: 1526–1532.

Sanderson, J. and M. Trolle. 2005. Monitoring elusive mammals. Unattended camera reveals secrets of some of the world's wildest places. American Scientist 93: 148–155.

Sayer, J.A., C.S. Harcourt and N.M. Collins (eds.). 1992. The Conservation Atlas of Tropical Forests. Africa. Simon and Schuster. New York.

Struhsaker, T.T. and M.I. Bakarr. 1999. A rapid survey of primates and other large mammals in Parc National de la Marahoué, Côte d'Ivoire. *In:* Schulenberg, T.S., C.A.

Short and P.J. Stephenson (eds.) A Biological Assessment of Parc National de la Marahoué. RAP Working Papers 13. Conservation International. Washington, DC. Pp. 50–53.

Waitkuwait, W.E. and J. Suter. (eds) 2001. Report on the establishment of a community-based bio-monitoring programme in and around Sapo National Park, Sinoe County, Liberia. Unpublished report Flora and Fauna International. Cambridge, UK.

Waitkuwait, W.E. and Suter, J. ed., 2003. Report on the First Year of Operation of a Comminuty-Based Bio-monitoring Programme in and around Sapo National Park, Sinoe County, Liberia. FFI, Cambridge, UK.

Whitesides, G.H., J.F. Oates, S.M. Green and R.P. Kluberdanz. 1988. Estimating primate densities from transects in a West African rain forest: a comparison of techniques. J. Anim. Ecol. 57: 345–367.

Wilkie, D.S., J.G. Sidle and G.C. Boundzanga. 1992. Mechanised logging, market hunting, and a bank loan in Congo. Conservation Biology 6: 570–580.

Appendix 1

Plant species recorded in North Lorma, Gola and Grebo National Forests.

Carel C.H. Jongkind

Herb. = Voucher deposited in National Herbarium Nederland - Wageningen University
-X- = Notes only taken in the field
Photo = Photograph and notes taken
Bold = Endemic to the Upper Guinea forest block
Bold & Underlined = Endemic to Liberia

Species' IUCN and CITES status are not listed because for western African plant species these lists are incomplete and represent only a small number of the plant species that are actually threatened. Combining the IUCN/CITES data with the list below would give a wrong indication for the sites we visited during this RAP survey.

Family	Species	North Lorma	Gola	Grebo
Acanthaceae	***Asystasia scandens* (Lindl.) Hook.**		Herb.	
Acanthaceae	*Asystasia vogeliana* Benth.		Herb.	
Acanthaceae	*Brillantaisia lamium* (Nees) Benth.	Herb.		
Acanthaceae	*Elytraria ivorensis* Dokosi			Herb.
Acanthaceae	*Elytraria marginata* Vahl	Herb.		
Acanthaceae	*Eremomastax speciosa* (Hochst.) Cufod.	Herb.		
Acanthaceae	*Justicia extensa* T.Anderson	Herb.		
Acanthaceae	*Justicia flava* (Forssk.) Vahl	Herb.		
Acanthaceae	*Justicia tenella* (Nees) T.Anderson	Herb.		
Acanthaceae	*Lankesteria brevior* C.B.Clarke		Herb.	
Acanthaceae	*Lepidagathis alopecuroides* (Vahl) R.Br. ex Griseb.	Herb.	Herb.	
Acanthaceae	***Mendoncia combretoides* (A.Chev.) Benoist**		Herb.	Herb.
Acanthaceae	*Physacanthus batanganus* (J.Braun & K.Schum.) Lindau		Herb.	
Acanthaceae	*Physacanthus nematosiphon* (Lindau) Rendle & Britten		Herb.	
Acanthaceae	*Rhinacanthus virens* (Nees) Milne-Redh.	Herb.		
Acanthaceae	*Ruellia primuloides* (T.Anderson ex Benth.) Heine	Herb.	Herb.	Herb.
Acanthaceae	***Staurogyne capitata* E.A.Bruce**	Herb.		Herb.
Acanthaceae	*Thunbergia chrysops* Hook.	Herb.		
Acanthaceae	***Whitfieldia colorata* C.B.Clarke ex Stapf**		Herb.	Herb.
Acanthaceae	***Whitfieldia lateritia* Hook.**	Herb.		Herb.
Adiantaceae	*Adiantum vogelii* Mett. ex Keyserl.	Herb.		Herb.
Adiantaceae	*Pellaea doniana* Hook.	Herb.		

continued

Family	Species	North Lorma	Gola	Grebo
Amaranthaceae	*Cyathula prostrata* (L.) Blume	Herb.		
Amaryllidaceae	*Crinum natans* Baker		-X-	
Anacardiaceae	*Trichoscypha arborea* (A.Chev.) A.Chev.			Herb.
Anacardiaceae	***Trichoscypha barbata* Breteler**		Herb.	
Anacardiaceae	*Trichoscypha bijuga* Engl.	Herb.	Herb.	
Anacardiaceae	***Trichoscypha linderi* Breteler**		Herb.	
Anacardiaceae	*Trichoscypha lucens* Oliv.			Herb.
Ancistrocladaceae	***Ancistrocladus barteri* Scott-Elliot**			Herb.
Anisophylleaceae	*Anisophyllea meniaudii* Aubrév. & Pellegr.		Herb.	
Annonaceae	*Annickia polycarpa* (DC.) Setten & Maas			Herb.
Annonaceae	*Artabotrys oliganthus* Engl. & Diels		Herb.	
Annonaceae	*Artabotrys* sp.	Herb.		
Annonaceae	*Cleistopholis patens* (Benth.) Engl. & Diels		Herb.	Herb.
Annonaceae	*Friesodielsia* sp.	Herb.		
Annonaceae	*Greenwayodendron oliveri* (Engl.) Verdc.	Herb.		
Annonaceae	*Monanthotaxis* sp. 1	Herb.		
Annonaceae	*Monanthotaxis* sp. 2	Herb.		
Annonaceae	*Monanthotaxis* sp. 3		Herb.	
Annonaceae	***Monocyclanthus vignei* Keay**		Herb.	
Annonaceae	*Monodora myristica* (Gaertn.) Dunal			Herb.
Annonaceae	*Neostenanthera gabonensis* (Engl. & Diels) Exell		Herb.	
Annonaceae	*Piptostigma fasciculatum* (De Wild.) Boutique	Herb.		Herb.
Annonaceae	***Piptostigma fugax* A.Chev. ex Hutch. & Dalziel**		Herb.	
Annonaceae	*Uvaria baumannii* Engl. & Diels			Herb.
Annonaceae	*Uvaria* sp. 1		Herb.	
Annonaceae	*Uvaria* sp. 2			Herb.
Annonaceae	*Uvaria* sp. 3			Herb.
Annonaceae	*Uvariastrum pierreanum* Engl. & Diels	Herb.		Herb.
Annonaceae	*Uvariopsis* sp.	Herb.		
Annonaceae	*Xylopia acutiflora* (Dunal) A.Rich.	Herb.	Herb.	
Annonaceae	*Xylopia le-testui* Pellegr.	Herb.		Herb.
Annonaceae	*Xylopia villosa* Chipp	Herb.		
Apocynaceae	*Alstonia boonei* De Wildeman		-X-	
Apocynaceae	*Ancylobotrys scandens* (Schumach. & Thonn.) Pichon			Herb.
Apocynaceae	*Baissea baillonii* Hua	Herb.		
Apocynaceae	*Callichilia subsessilis* (Benth.) Stapf	Herb.	Herb.	
Apocynaceae	***Hunteria simii* (Stapf) H.Huber**	Herb.	Herb.	Herb.
Apocynaceae	*Landolphia dulcis* (R.Br. ex Sabine) Pichon		Herb.	
Apocynaceae	*Landolphia incerta* (K.Schum.) J.G.M.Pers.			Herb.
Apocynaceae	***Landolphia nitidula* J.G.M.Pers.**			Herb.
Apocynaceae	*Landolphia owariensis* P.Beauv.			Herb.
Apocynaceae	*Oncinotis gracilis* Stapf	Herb.		
Apocynaceae	*Orthopichonia* sp.	Herb.		
Apocynaceae	*Pleiocarpa mutica* Benth.		Herb.	

continued

Family	Species	North Lorma	Gola	Grebo
Apocynaceae	*Tabernaemontana psorocarpa* (Pierre ex Stapf) Pichon		Herb.	
Araceae	*Amorphophallus* sp.	-X-		
Araceae	***Anubias gigantea* A.Chev. ex Hutch.**	Herb.		
Araceae	*Anubias gracilis* A.Chev. ex Hutch.	Herb.	Herb.	
Araceae	*Cercestis afzelii* Schott	Herb.		-X-
Araceae	*Cercestis dinklagei* Engl.	Herb.		Herb.
Araceae	*Cercestis* sp.	Herb.		
Araceae	*Culcasia angolensis* Welw. ex Schott	Herb.		
Araceae	*Culcasia sapinii* De Wild.	Herb.		Herb.
Araceae	*Culcasia scandens* P.Beauv.	Herb.		Herb.
Araceae	*Rhaphidophora africana* N.E.Br.	Herb.		
Asclepiadaceae	*Periploca nigrescens* Afzel.			Herb.
Asclepiadaceae	*Tylophora cuspidata* (K.Schum.) Meve & Omlor			Herb.
Aspleniaceae	*Asplenium africanum* Desv.	Herb.	Herb.	
Aspleniaceae	*Asplenium anisophyllum* Kunze	Herb.		
Aspleniaceae	*Asplenium barteri* Hook.	Herb.	Herb.	
Aspleniaceae	*Asplenium formosum* Willd.	Herb.		
Aspleniaceae	*Asplenium unilaterale* Lam.	Herb.		
Aspleniaceae	*Asplenium variabile* Hook.		Herb.	Herb.
Begoniaceae	***Begonia cavallyensis* A.Chev.**		Herb.	
Begoniaceae	*Begonia fusialata* Warb. var. *fusialata*	Herb.		Herb.
Begoniaceae	*Begonia polygonoides* Hook.f.	Herb.		
Begoniaceae	*Begonia quadrialata* Warb. subsp. *quadrialata*	Herb.		
Bignoniaceae	*Newbouldia laevis* (P.Beauv.) Seeman ex Bureau	Herb.		
Bombacaceae	*Ceiba pentandra* (L.) Gaertn.	-X-		
Burmanniaceae	*Burmannia congesta* (Wright) Jonker		Herb.	
Burmanniaceae	*Gymnosiphon longistylus* (Benth.) Hutch.	Herb.	Herb.	
Burseraceae	*Dacryodes klaineana* (Pierre) H.J.Lam		Herb.	-X-
Capparaceae	*Euadenia eminens* Hook.f.			Herb.
Capparaceae	*Ritchiea capparoides* (Andr.) Britten		Herb.	Herb.
Celastraceae	*Salacia lehmbachii* Loes.		Herb.	
Celastraceae	*Salacia owabiensis* Hoyle			Herb.
Celastraceae	*Salacia* sp.		Herb.	
Celastraceae	*Salacia staudtiana* Loes.	Herb.		
Chrysobalanaceae	*Afrolicania elaeosperma* Mildbr.	Herb.		
Chrysobalanaceae	***Dactyladenia hirsuta* (A.Chev. & De Wild.) Prance & F.White**		Herb.	
Chrysobalanaceae	***Dactyladenia whytei* (Stapf) Prance & White**		Herb.	
Chrysobalanaceae	*Magnistipula zenkeri* Engl.			Herb.
Chrysobalanaceae	***Maranthes aubrevillei* (Pellegr.) Prance**	Herb.		Herb.
Chrysobalanaceae	*Maranthes glabra* (Oliv.) Prance	Herb.		
Chrysobalanaceae	*Parinari excelsa* Sabine	Herb.		Herb.
Combretaceae	*Combretum aphanopetalum* Engl. & Diels			Herb.
Combretaceae	*Combretum comosum* G.Don			Herb.

continued

Family	Species	North Lorma	Gola	Grebo
Combretaceae	*Combretum oyemense* Exell	Herb.		
Combretaceae	***Strephonema pseudocola* A.Chev.**	Herb.	Herb.	Herb.
Combretaceae	*Terminalia ivorensis* A.Chevalier	-X-		
Combretaceae	*Terminalia superba* Engler & Diels			-X-
Commelinaceae	***Buforrestia obovata* Brenan**	Herb.		
Commelinaceae	*Commelina capitata* Benth.	Herb.	Herb.	Herb.
Commelinaceae	*Floscopa africana* (P.Beauv.) C.B.Clarke		Herb.	Herb.
Commelinaceae	*Palisota bracteosa* C.B.Clarke	Herb.		Herb.
Commelinaceae	*Pollia condensata* C.B.Clarke	Herb.		Herb.
Commelinaceae	*Polyspatha paniculata* Benth.	Herb.		
Commelinaceae	*Stanfieldiella imperforata* (C.B.Clarke) Brenan	Herb.		
Compositae	*Adenostemma perrottetii* DC.		Herb.	
Compositae	*Chromolaena odorata* (L.) R.M.King & H.Rob.	Herb.		
Compositae	*Vernonia titanophylla* Brenan	-X-		
Connaraceae	*Agelaea paradoxa* Gilg var. *microcarpa* Jongkind	-X-		Herb.
Connaraceae	*Agelaea pentagyna* (Lam.) Baill.	-X-	Herb.	Herb.
Connaraceae	***Cnestis bomiensis* Lemmens**		Herb.	
Connaraceae	*Connarus africanus* Lam.	Herb.		
Connaraceae	*Manotes expansa* Sol. ex Planchon		-X-	
Connaraceae	*Manotes macrantha* (Gilg) Schellenb.		Herb.	
Connaraceae	*Rourea minor* (Gaertn.) Alston		Herb.	Herb.
Connaraceae	*Rourea solanderi* Baker		Herb.	Herb.
Connaraceae	*Rourea thomsonii* (Baker) Jongkind			Herb.
Convolvulaceae	*Bonamia thunbergiana* (Roem. & Schult.) F.N.Williams			Herb.
Convolvulaceae	*Calycobolus africanus* (G.Don) Heine	Herb.		
Convolvulaceae	*Calycobolus heudelotii* (Baker ex Oliv.) Heine	Herb.		
Convolvulaceae	*Ipomoea aitonii* Lindl.	Herb.		
Convolvulaceae	*Ipomoea obscura* (L.) Ker Gawl.	Herb.		
Convolvulaceae	*Neuropeltis acuminata* (P.Beauv.) Benth.			Herb.
Convolvulaceae	*Stictocardia beraviensis* (Vatke) Hallier f.	Herb.		
Costaceae	***Costus deistelii* K.Schum.**	Herb.		
Costaceae	*Costus* sp.	-X-	-X-	
Cucurbitaceae	*Momordica charantia* L.	Herb.		
Cucurbitaceae	*Momordica foetida* Schumach.			Herb.
Cyatheaceae	*Cyathea camerooniana* Hook.	Herb.		
Cyperaceae	*Hypolytrum heteromorphum* Nelmes	Herb.		
Cyperaceae	*Hypolytrum purpurascens* Cherm.	Herb.		
Cyperaceae	*Hypolytrum* sp. 1	Herb.		
Cyperaceae	*Hypolytrum* sp. 2			Herb.
Cyperaceae	***Mapania ivorensis* (Raynal) Raynal**		Herb.	
Cyperaceae	***Mapania linderi* Hutch. ex Nelmes**	Herb.	Herb.	
Cyperaceae	*Scleria boivinii* Steud.		-X-	
Cyperaceae	*Scleria naumanniana* Boeckeler		Herb.	

continued

Family	Species	North Lorma	Gola	Grebo
Dennstaedtiaceae	*Microlepia speluncae* (L.) Moore			Herb.
Dichapetalaceae	*Dichapetalum angolense* Chodat			Herb.
Dichapetalaceae	*Dichapetalum heudelotii* (Planch. ex Oliv.) Baill.		Herb.	
Dichapetalaceae	*Dichapetalum* sp.			Herb.
Dichapetalaceae	***Dichapetalum toxicarium* (G.Don) Baill.**			Herb.
Dilleniaceae	*Tetracera alnifolia* Willd.		-X-	-X-
Dioncophyllaceae	***Triphyophyllum peltatum* (Hutch. & Dalziel) Airy Shaw**		Herb.	Herb.
Dioscoreaceae	*Dioscorea* sp.	Herb.		
Dracaenaceae	*Dracaena aubryana* Brongn. ex C.J.Morren	Herb.	-X-	Herb.
Dracaenaceae	*Dracaena camerooniana* Baker			Herb.
Dracaenaceae	***Dracaena cristula* W.Bull**			Herb.
Dracaenaceae	*Dracaena ovata* Ker Gawl.		Herb.	
Dracaenaceae	*Dracaena surculosa* Lindl. var. *maculata* Hook.f.			Herb.
Dracaenaceae	*Sansevieria liberica* Gér. & Labr.	-X-		
Dryopteridaceae	*Callipteris prolifera* (Lam.) Bory			Herb.
Dryopteridaceae	*Tectaria* sp.	Herb.		
Dryopteridaceae	*Triplophyllum buchholzii* (Kuhn) Holttum		Herb.	
Dryopteridaceae	*Triplophyllum* sp. 1	Herb.		
Dryopteridaceae	*Triplophyllum* sp. 2		Herb.	
Dryopteridaceae	*Triplophyllum* sp. 3		Herb.	
Ebenaceae	***Diospyros chevalieri* De Wild.**		Herb.	Herb.
Ebenaceae	*Diospyros ferrea* (Willd.) Bakh.		Herb.	
Ebenaceae	*Diospyros gabunensis* Gürke		Herb.	
Ebenaceae	***Diospyros heudelotii* Hiern**	Herb.		
Ebenaceae	*Diospyros mannii* Hiern	Herb.	Herb.	Herb.
Ebenaceae	*Diospyros sanza-minika* A.Chev.			Herb.
Ebenaceae	*Diospyros soubreana* F.White			Herb.
Ebenaceae	*Diospyros* sp.	Herb.		
Erythroxylaceae	*Erythroxylum mannii* Oliv.			Herb.
Euphorbiaceae	*Alchornea cordifolia* (Schum. & Thonning) Muell.Arg.		-X-	-X-
Euphorbiaceae	*Antidesma* sp.			Herb.
Euphorbiaceae	***Crotonogyne caterviflora* N.E.Br.**		Herb.	
Euphorbiaceae	*Discoglypremna caloneura* (Pax) Prain			Herb.
Euphorbiaceae	***Macaranga heterophylla* (Muell.Arg.) Muell.Arg.**	-X-	-X-	
Euphorbiaceae	*Macaranga hurifolia* Beille			-X-
Euphorbiaceae	*Maesobotrya barteri* (Baillon) Hutch.	-X-		
Euphorbiaceae	*Manniophyton fulvum* Müll.Arg.	-X-	-X-	-X-
Euphorbiaceae	*Mareya micrantha* (Benth.) Müll.Arg.		Herb.	
Euphorbiaceae	***Phyllanthus kerstingii* Brunel**	Herb.		
Euphorbiaceae	*Phyllanthus profusus* N.E.Br.	Herb.		
Euphorbiaceae	*Plesiatropha paniculata* (Pax) Breteler	Herb.		
Euphorbiaceae	*Spondianthus preussii* Engler			-X-
Euphorbiaceae	*Tragia spathulata* Benth.	Herb.		

continued

Family	Species	North Lorma	Gola	Grebo
Euphorbiaceae	*Uapaca paludosa* Aubrév. & Léandri			Herb.
Flacourtiaceae	*Oncoba brevipes* Stapf		-X-	
Flacourtiaceae	*Oncoba echinata* Oliv.	Herb.		
Gentianaceae	*Anthocleista nobilis* G.Don		-X-	
Gentianaceae	*Voyria primuloides* Baker		Herb.	
Gleicheniaceae	*Dicranopteris linearis* (Burm.) Underwood		-X-	-X-
Gramineae	*Acroceras gabunense* (Hack.) Clayton	Herb.		
Gramineae	*Centotheca lappacea* (L.) Desv.	Herb.	Herb.	
Gramineae	*Guaduella oblonga* Hutch. ex W.D.Clayton	Herb.	Herb.	
Gramineae	*Leptaspis zeylanica* Nees ex Steud.	Herb.		
Gramineae	*Olyra latifolia* L.	Herb.		
Gramineae	*Oplismenus hirtellus* (L.) P.Beauv.	Herb.		
Gramineae	*Panicum laxum* Sw.		Herb.	
Gramineae	*Pseudechinolaena polystachya* (Kunth) Stapf	Herb.		
Gramineae	*Setaria megaphylla* (Steud.) Dur. & Schinz	Herb.		
Gramineae	*Streptogyna crinita* P.Beauv.	Herb.		Herb.
Grammitidaceae	*Cochlidium serrulatum* (Swartz) L.E. Bishop		Herb.	
Guttiferae	*Garcinia epunctata* Stapf			Herb.
Guttiferae	*Harungana madagascariensis* Lamarck ex Poiret		-X-	-X-
Guttiferae	*Mammea africana* Sabine			-X-
Guttiferae	*Pentadesma butyracea* Sabine	-X-	-X-	-X-
Humiriaceae	*Sacoglottis gabonensis* (Baill.) Urb.		Herb.	Herb.
Hymenophyllaceae	*Hymenophyllum hirsutum* (L.) Sw.		Herb.	
Hymenophyllaceae	*Trichomanes chamaedrys* Taton		Herb.	
Hymenophyllaceae	*Trichomanes fallax* Christ	Herb.		
Hymenophyllaceae	*Trichomanes guineense* Afzel. ex Sw.		Herb.	
Icacinaceae	*Desmostachys vogelii* (Miers) Stapf		Herb.	
Icacinaceae	***Iodes liberica* Stapf**			Herb.
Icacinaceae	*Pyrenacantha acuminata* Engl.	Herb.		
Icacinaceae	*Pyrenacantha glabrescens* (Engl.) Engl.			Herb.
Icacinaceae	*Pyrenacantha klaineana* Pierre ex Exell & Mendonça		Herb.	
Icacinaceae	***Rhaphiostylis cordifolia* Hutch. & Dalziel**		Herb.	
Icacinaceae	*Rhaphiostylis* sp. nov.		Herb.	Herb.
Irvingiaceae	*Irvingia gabonensis* (Aubry-Lecomte) Baillon			-X-
Labiatae	*Achyrospermum oblongifolium* Baker	Herb.		Herb.
Labiatae	*Plectranthus epilithicus* B.J.Pollard	Herb.		
Labiatae	*Plectranthus* sp.	Herb.		
Lecythidaceae	*Napoleonaea vogelii* Hook. & Planch.			Herb.
Lecythidaceae	*Petersianthus macrocarpus* (P.Beauv.) Liben	-X-		
Leguminosae-Caes.	*Afzelia* sp.		-X-	
Leguminosae-Caes.	*Anthonotha crassifolia* (Baill.) J.Léonard	Herb.		
Leguminosae-Caes.	*Anthonotha fragrans* (Baker f.) Exell & Hillcoat	Herb.	Herb.	
Leguminosae-Caes.	***Bussea occidentalis* Hutch. ex Chipp.**	Herb.		

continued

Family	Species	North Lorma	Gola	Grebo
Leguminosae-Caes.	***Copaifera salikounda* Heckel**	Herb.	-X-	Herb.
Leguminosae-Caes.	***Cryptosepalum tetraphyllum* Benth.**	Herb.		
Leguminosae-Caes.	***Daniella thurifera* Bennett**	-X-		
Leguminosae-Caes.	***Dialium aubrevillei* Pellegr.**	Herb.	Herb.	Herb.
Leguminosae-Caes.	*Distemonanthus benthamianus* Baillon	-X-		
Leguminosae-Caes.	*Erythrophleum ivorense* A.Chev.			Herb.
Leguminosae-Caes.	***Gilbertiodendron aylmeri* (Hutch. & Dalziel) J.Léonard**		Herb.	
Leguminosae-Caes.	*Gilbertiodendron preussii* (Harms) J.Léonard	-X	Herb.	Herb.
Leguminosae-Caes.	*Griffonia simplicifolia* (Vahl ex DC.) Baillon			-X-
Leguminosae-Caes.	***Guibourtia leonensis* J.Léonard**	Herb.		
Leguminosae-Caes.	*Paramacrolobium coeruleum* (Taub.) J.Léonard	Herb.		
Leguminosae-Caes.	*Plagiosiphon emarginatus* (Hutch. & Dalziel) J. Léonard	Herb.	Herb.	
Leguminosae-Caes.	*Senna podocarpa* (Guill. & Perr.) Lock	Herb.		
Leguminosae-Caes.	*Senna tora* (L.) Roxb.	Herb.		
Leguminosae-Caes.	***Stachyothyrsus stapfiana* (A.Chev.) J.Léonard & Voorhoeve**		Herb.	
Leguminosae-Caes.	***Tessmannia baikiaeoides* Hutch. & Dalziel**	Herb.		
Leguminosae-Mim.	***Calpocalyx brevibracteatus* Harms**		Herb.	
Leguminosae-Mim.	*Newtonia duparquetiana* (Baill.) Keay	Herb.		
Leguminosae-Mim.	*Newtonia* sp.			Herb.
Leguminosae-Mim.	*Parkia bicolor* A.Chev.	Herb.	-X-	
Leguminosae-Mim.	*Pentaclethra macrophylla* Bentham	-X-		-X-
Leguminosae-Mim.	*Piptadeniastrum africanum* (Hooker f.) Brenan			-X-
Leguminosae-Mim.	***Xylia evansii* Hutch.**	Herb.		
Leguminosae-Pap.	*Abrus fruticulosus* Wall. ex W. & A.	Herb.		
Leguminosae-Pap.	*Amphimas pterocarpoides* Harms			Herb.
Leguminosae-Pap.	*Baphia capparidifolia* Baker subsp. *polygalacea* Brummitt	Herb.		
Leguminosae-Pap.	*Baphia nitida* Lodd.			Herb.
Leguminosae-Pap.	*Dalbergia adamii* Berhaut	Herb.		
Leguminosae-Pap.	*Dalbergia afzeliana* G.Don			Herb.
Leguminosae-Pap.	*Dalbergia heudelotii* Stapf	Herb.		
Leguminosae-Pap.	*Dalbergia oblongifolia* G.Don	Herb.		Herb.
Leguminosae-Pap.	*Dalbergia* sp.		Herb.	
Leguminosae-Pap.	*Leptoderris sassandrensis* Jongkind			Herb.
Leguminosae-Pap.	*Leptoderris* sp. nov.			Herb.
Leguminosae-Pap.	*Millettia chrysophylla* Dunn	Herb.		Herb.
Leguminosae-Pap.	***Millettia lane-poolei* Dunn**	Herb.		
Leguminosae-Pap.	***Millettia liberica* Jongkind**			Herb.
Leguminosae-Pap.	***Millettia lucens* (Scott-Elliot) Dunn**	Herb.		
Leguminosae-Pap.	*Millettia* sp.	Herb.		
Leguminosae-Pap.	***Millettia warneckei* Harms var. *porphyrocalyx* (Dunn) Hepper**	Herb.		
Leguminosae-Pap.	***Platysepalum hirsutum* (Dunn) Hepper**			Herb.
Leguminosae-Pap.	*Vigna gracilis* (Guill. & Perr.) Hook.f.	Herb.		Herb.
Liliaceae	*Asparagus drepanophyllus* Welw.	Herb.		

continued

Family	Species	North Lorma	Gola	Grebo
Liliaceae	*Chlorophytum alismaefolium* Baker	Herb.		
Liliaceae	*Chlorophytum comosum* (Thunb.) Jacq. var. *sparsiflorum* (Baker) A.D.Poulsen & Nordal	Herb.	Herb.	
Linaceae	*Hugonia* sp.	-X-		
Linaceae	*Ochthocosmus africanus* Hook.f.	Herb.		
Loganiaceae	*Strychnos aculeata* Solereder		-X-	
Loganiaceae	*Strychnos afzelii* Gilg	Herb.		Herb.
Loganiaceae	*Strychnos barteri* Soler.			Herb.
Loganiaceae	*Strychnos camptoneura* Gilg & Busse		Herb.	
Loganiaceae	*Strychnos densiflora* Baill.		Herb.	
Loganiaceae	*Strychnos icaja* Baill.	Herb.		Herb.
Loganiaceae	*Strychnos splendens* Gilg	Herb.		
Loganiaceae	*Strychnos usambarensis* Gilg		Herb.	
Lomariopsidaceae	*Bolbitis acrostichoides* (Afzel. ex Sw.) Ching	Herb.		
Lomariopsidaceae	*Bolbitis salicina* (Hook.) Ching	Herb.	Herb.	
Lomariopsidaceae	*Bolbitis* sp.		Herb.	
Lomariopsidaceae	*Elaphoglossum* sp. 1		Herb.	
Lomariopsidaceae	*Elaphoglossum* sp. 2		Herb.	
Lomariopsidaceae	*Lomariopsis guineensis* (Underw.) Alston		Herb.	Herb.
Lomariopsidaceae	*Lomariopsis palustris* (Hook.) Mett. ex Kuhn	Herb.		Herb.
Loxogrammataceae	*Loxogramme abyssinica* (Baker) M.G.Price	Herb.		
Lycopodiaceae	*Lycopodiella cernua* (L.) Pichi Sermolli			Photo
Malpighiaceae	***Acridocarpus longifolius* (G.Don) Hook.f.**			Herb.
Malpighiaceae	***Acridocarpus plagiopterus* Guill. & Perr.**			Herb.
Malpighiaceae	*Flabellaria paniculata* Cav.	Herb.		
Malvaceae	*Wissadula amplissima* (L.) R.E.Fr. var. *rostrata* (Schumach. & Thonn.) R.E.Fr.	Herb.		
Marantaceae	*Halopegia azurea* (K.Schum.) K.Schum.	Herb.	Herb.	
Marantaceae	***Marantochloa cuspidata* (Rosc.) Milne-Redh.**	Herb.		
Marantaceae	*Marantochloa filipes* (Benth.) Hutch.	Herb.		
Marantaceae	*Marantochloa leucantha* (K.Schum.) Milne-Redh.			Herb.
Marantaceae	*Sarcophrynium brachystachyum* (Benth.) K.Schum.		Herb.	Herb.
Marattiaceae	*Marattia fraxinea* J.Sm.		Herb.	
Medusandraceae	***Soyauxia floribunda* Hutch.**	Herb.		
Melastomataceae	*Calvoa monticola* A.Chev. ex Hutch. & Dalziel	Herb.		
Melastomataceae	*Dicellandra barteri* Hook.f.		Herb.	
Melastomataceae	*Dichaetanthera africana* (Hook.f.) Jacq.-Fél.			Herb.
Melastomataceae	*Guyonia ciliata* Hook.f.		Herb.	
Melastomataceae	*Melastomastrum theifolium* (G.Don) A.Fern. & R.Fern.	Herb.		
Melastomataceae	*Memecylon lateriflorum* (G.Don) Bremek.		Herb.	
Melastomataceae	*Memecylon* sp.	Herb.		
Melastomataceae	*Ochthocharis dicellandroides* (Gilg) C.Hansen & Wickens			Herb.
Melastomataceae	*Tristemma akeassii* Jacq.-Fél.			Herb.
Melastomataceae	***Tristemma coronatum* Benth.**	Herb.	Herb.	

continued

Family	Species	North Lorma	Gola	Grebo
Melastomataceae	*Warneckea cinnamomoides* (G.Don) Jacq.-Fél.		Herb.	
Melastomataceae	***Warneckea golaensis* (Baker f.) Jacq.-Fél.**	Herb.		
Melastomataceae	*Warneckea memecyloides* (Benth.) Jacq.-Fél.			Herb.
Melastomataceae	*Warneckea* sp.		Herb.	
Meliaceae	*Carapa procera* DC.	Herb.		
Meliaceae	*Entandrophragma angolensis* (Welwitsch) DC.	-X-		
Meliaceae	*Entandrophragma utile* (Dawe & Sprague) Sprague	-X-		
Meliaceae	*Khaya* sp.			-X-
Menispermaceae	***Albertisia ferruginea* (Diels) Forman**		Herb.	
Menispermaceae	***Kolobopetalum leonense* Hutch. & Dalziel**		Herb.	
Menispermaceae	***Penianthus patulinervis* Hutch. & Dalziel**	Herb.		
Menispermaceae	***Tiliacora leonensis* (Scott-Elliot) Diels**			Herb.
Moraceae	*Antiaris toxicaria* (Rumph. ex Pers.) Leschen.			-X-
Moraceae	*Ficus barteri* Sprague			Herb.
Moraceae	*Ficus elasticoides* De Wild.			Herb.
Moraceae	*Ficus leonensis* Hutch.		Herb.	
Moraceae	*Ficus lingua* Warb. ex De Wild. & T.Durand subsp. *lingua*			Herb.
Moraceae	*Ficus natalensis* Hochst. subsp. *leprieurii* (Miq.) C.C.Berg	Herb.		
Moraceae	***Ficus pachyneura* C.C.Berg**		Herb.	
Moraceae	*Ficus sansibarica* Warb.	Herb.		
Moraceae	*Ficus saussureana* DC.		Herb.	
Moraceae	*Ficus umbellata* Vahl			Herb.
Moraceae	*Ficus vogeliana* (Miq.) Miq.			Herb.
Moraceae	***Milicia regia* (A.Chev.) C.C.Berg**		Herb.	
Moraceae	*Musanga cecropioides* F.Br.	-X-	-X-	-X-
Moraceae	***Myrianthus libericus* Rendle**	Herb.		
Moraceae	*Streblus usambarensis* (Engl.) C.C.Berg	Herb.		
Moraceae	*Treculia africana* Decne.	Herb.		Herb.
Moraceae	*Trilepisium madagascariense* Thouars ex DC.	Herb.		
Myristicaceae	*Pycnanthus angolensis* (Welwitsch) Warb.	-X-		-X-
Ochnaceae	***Campylospermum amplectens* (Stapf) Farron**		Herb.	
Ochnaceae	*Campylospermum congestum* (Oliv.) Farron	Herb.		
Ochnaceae	*Campylospermum duparquetianum* (Baill.) Tiegh.			Herb.
Ochnaceae	*Campylospermum glaberrimum* (P.Beauv.) Farron		Herb.	
Ochnaceae	*Campylospermum schoenleinianum* (Klotzsch) Farron	Herb.	Herb.	Herb.
Ochnaceae	***Campylospermum subcordatum* (Stapf) Farron**	Herb.	Herb.	
Ochnaceae	*Lophira alata* Banks ex Gaertn.		-X-	-X-
Ochnaceae	*Ochna membranacea* Oliv.	Herb.		
Ochnaceae	*Rhabdophyllum calophyllum* (Hook.f.) Tiegh.	Herb.		
Olacaceae	*Coula edulis* Baill.			Herb.
Olacaceae	*Heisteria parvifolia* Sm.		-X-	Herb.
Olacaceae	*Olax gambecola* Baill.	Herb.		
Olacaceae	***Ptychopetalum anceps* Oliv.**		Herb.	Herb.

continued

Family	Species	North Lorma	Gola	Grebo
Olacaceae	*Strombosia pustulata* Oliv.		Herb.	Herb.
Oleaceae	*Jasminum pauciflorum* Benth.	Herb.		
Oleandraceae	*Arthropteris palisotii* (Desv.) Alston			Herb.
Oleandraceae	*Nephrolepis biserrata* (Sw.) Schott	Herb.		
Orchidaceae	*Angraecum birrimense* Rolfe	Herb.		
Orchidaceae	*Angraecum distichum* Lindl.		Herb.	
Orchidaceae	*Angraecum podochiloides* Schltr.		Herb.	
Orchidaceae	*Angraecum subulatum* Lindl.	Herb.		
Orchidaceae	*Bulbophyllum magnibracteatum* Summerh.		Herb.	
Orchidaceae	*Bulbophyllum oreonastes* Rchb.f.		Herb.	
Orchidaceae	*Calyptrochilum christyanum* (Rchb.f.) Summerh.			Herb.
Orchidaceae	*Chamaeangis odoratissima* (Rchb.f.) Schltr.			Herb.
Orchidaceae	*Habenaria macrandra* Lindl.	Herb.		
Orchidaceae	*Nervilia* sp.	Photo		
Orchidaceae	*Oeceoclades maculata* (Lindley) Lindley	Photo		
Orchidaceae	*Polystachya* Hook.			Herb.
Orchidaceae	*Polystachya polychaete* Kraenzl.		Herb.	
Orchidaceae	*Tridactyle bicaudata* (Lindl.) Schltr.		Herb.	
Orchidaceae	*Vanilla africana* Lindl.	Herb.		
Palmae	*Eremospatha* sp.		-X-	
Palmae	*Laccosperma* sp.		-X-	
Palmae	*Raphia hookeri* Mann & Wendl.			-X-
Palmae	***Raphia palma-pinus* (Gaertn.) Hutch.**	-X-		
Pandaceae	*Microdesmis keayana* J.Léonard	Herb.		-X-
Pandaceae	*Panda oleosa* Pierre			Herb.
Passifloraceae	*Adenia cissampeloides* (Planch. ex Benth.) Harms			Herb.
Passifloraceae	*Adenia lobata* (Jacq.) Engler			-X-
Passifloraceae	*Adenia mannii* (Mast.) Engl.			Herb.
Passifloraceae	***Androsiphonia adenostegia* Stapf**		Herb.	
Passifloraceae	***Crossostemma laurifolium* Planch. ex Benth.**	Herb.		
Passifloraceae	*Smeathmannia pubescens* Sol. ex R.Br.	Herb.		
Piperaceae	*Peperomia rotundifolia* (L.) H.B.& K.	Herb.		
Piperaceae	*Piper guineense* Schum. & Thonning	-X-	-X-	-X-
Piperaceae	*Piper umbellatum* L.			-X-
Polygalaceae	*Carpolobia alba* G.Don	Herb.		
Polygonaceae	*Afrobrunnichia erecta* (Asch.) Hutch. & Dalziel	-X-		Herb.
Polypodiaceae	*Drynaria laurentii* (Christ) Hieronymus			-X-
Polypodiaceae	*Microgramma lycopodioides* (L.) Copel.			Herb.
Polypodiaceae	*Microsorium punctatum* (L.) Copeland	-X-		-X-
Polypodiaceae	*Phymatosorus scolopendria* (Burm.f.) Pic.Serm.			Herb.
Polypodiaceae	*Platycerium stemaria* (P.Beauv.) Desvaux	-X-		Photo
Pteridaceae	*Pityrogramma calomelanos* (L.) Link			-X-
Pteridaceae	*Pteris burtonii* Baker		Herb.	

continued

Family	Species	North Lorma	Gola	Grebo
Putranjivaceae	*Drypetes gilgiana* (Pax) Pax & K.Hoffm.	Herb.	Herb.	
Putranjivaceae	*Drypetes inaequalis* Hutch.	Herb.		
Putranjivaceae	***Drypetes* sp. nov.**			Herb.
Rapateaceae	***Maschalocephalus dinklagei* Gilg & K.Schum.**		Herb.	
Rhamnaceae	*Lasiodiscus fasciculiflorus* Engl.	Herb.	Herb.	
Rhamnaceae	*Lasiodiscus mannii* Hook.f.		Herb.	
Rhamnaceae	*Ventilago africana* Exell			Herb.
Rhizophoraceae	***Cassipourea nialatou* Aubrév. & Pellegr.**			Herb.
Rubiaceae	***Argocoffeopsis afzelii* (Hiern) Robbr.**	Herb.		
Rubiaceae	***Argostemma pumilum* Benn.**		Herb.	
Rubiaceae	*Bertiera bracteolata* Hiern			Herb.
Rubiaceae	*Bertiera breviflora* Hiern.	Herb.		
Rubiaceae	*Bertiera racemosa* (G.Don) K.Schum.			Herb.
Rubiaceae	***Bertiera spicata* (C.F.Gaertn.) K.Schum.**		Herb.	
Rubiaceae	***Cephaelis micheliae* J.-G.Adam**		Herb.	
Rubiaceae	*Chassalia afzelii* (Hiern) K.Schum.			Herb.
Rubiaceae	***Chassalia corallifera* (A.Chev. ex De Wild.) Hepper**		Herb.	
Rubiaceae	***Chassalia* sp. nov.**		Herb.	
Rubiaceae	*Corynanthe pachyceras* K.Schum.			Herb.
Rubiaceae	*Craterispermum caudatum* Hutch.			Herb.
Rubiaceae	*Cremaspora triflora* (Thonn.) K.Schum.			Herb.
Rubiaceae	*Gaertnera longevaginalis* (Hiern) E.M.A.Petit	Herb.		
Rubiaceae	*Gaertnera* sp.		Herb.	
Rubiaceae	*Gardenia nitida* Hook.	Herb.		
Rubiaceae	*Geophila afzelii* Hiern	Herb.		Herb.
Rubiaceae	*Geophila obvallata* (Schumach.) F.Didr.	Herb.		
Rubiaceae	*Heinsia crinita* (Afzel.) G.Taylor	-X-	Herb.	-X-
Rubiaceae	*Hutchinsonia barbata* Robyns		Herb.	
Rubiaceae	*Hymenocoleus neurodictyon* (K.Schum.) Robbr.	Herb.		
Rubiaceae	*Hymenocoleus* sp. 1		Herb.	
Rubiaceae	*Hymenocoleus* sp. 2		Herb.	
Rubiaceae	***Ixora aggregata* Hutch.**			Herb.
Rubiaceae	***Ixora nimbana* Schnell**	Herb.		
Rubiaceae	***Keetia bridsoniae* Jongkind**	Herb.		
Rubiaceae	*Keetia leucantha* (Krause) Bridson			Herb.
Rubiaceae	***Keetia obovata* Jongkind**		Herb.	
Rubiaceae	*Keetia rufivillosa* (Robyns ex Hutch. & Dalziel) Bridson	Herb.		Herb.
Rubiaceae	*Keetia* sp.		Herb.	
Rubiaceae	*Lasianthus batangensis* K.Schum.	Herb.	Herb.	
Rubiaceae	*Lasianthus repens* Hepper	Herb.		
Rubiaceae	*Massularia acuminata* (G.Don) Bullock ex Hoyle	Herb.		
Rubiaceae	***Mussaenda chippii* Wernham**		Herb.	
Rubiaceae	***Mussaenda grandiflora* Benth.**			Herb.

continued

Family	Species	North Lorma	Gola	Grebo
Rubiaceae	*Nauclea diderrichii* (De Wild. & Th.Dur.) Merrill	-X-		-X-
Rubiaceae	*Nauclea vanderguchtii* (De Wild.) Petit		Herb.	
Rubiaceae	*Nichallea soyauxii* (Hiern) Bridson			Herb.
Rubiaceae	*Oxyanthus formosus* Hook.f. ex Planch.	Herb.		
Rubiaceae	*Parapentas setigera* (Hiern) Verdc.	Herb.		
Rubiaceae	*Pauridiantha sylvicola* (Hutch. & Dalziel) Bremek.	Herb.	Herb.	Herb.
Rubiaceae	*Pavetta* sp.		Herb.	
Rubiaceae	*Poecilocalyx stipulosa* (Hutch. & Dalziel) N.Hallé		Herb.	
Rubiaceae	*Psychotria biaurita* (Hutch. & Dalziel) Verdc.	Herb.		
Rubiaceae	*Psychotria gabonica* Hiern			Herb.
Rubiaceae	***Psychotria kwewonii* Jongkind ined.**			Herb.
Rubiaceae	***Psychotria ombrophila* (Schnell) Verdc.**		Herb.	
Rubiaceae	*Psychotria peduncularis* (Salisb.) Verdcourt			-X-
Rubiaceae	*Psychotria* sp. 1	Herb.		
Rubiaceae	*Psychotria* sp. 2			Herb.
Rubiaceae	*Psychotria* sp. 3			Herb.
Rubiaceae	*Psychotria* sp. 4			Herb.
Rubiaceae	***Psychotria yapoensis* (Schnell) Verdc.**			Herb.
Rubiaceae	*Rothmannia whitfieldii* (Lindl.) Dandy		Herb.	
Rubiaceae	*Rytigynia canthioides* (Benth.) Robyns			Herb.
Rubiaceae	***Sabicea ferruginea* Benth.**		-X-	Herb.
Rubiaceae	*Sabicea rosea* Hoyle			Herb.
Rubiaceae	***Schizocolea linderi* (Hutch. & Dalziel) Bremek.**	Herb.	Herb.	
Rubiaceae	***Sericanthe adamii* (N.Hallé) Robbr.**		Herb.	
Rubiaceae	***Sherbournia calycina* (G.Don) Hua**			Herb.
Rubiaceae	***Stelechantha ziamaeana* (Jacq.-Fél.) N.Hallé**	Herb.	Herb.	Herb.
Rubiaceae	*Tarenna fusco-flava* (K.Schum.) S.Moore			Herb.
Rubiaceae	*Tarenna* sp.		Herb.	
Rubiaceae	*Tricalysia pallens* Hiern			Herb.
Rubiaceae	*Tricalysia reflexa* Hutch.		Herb.	
Rubiaceae	*Tricalysia* sp. nov.		Herb.	
Rubiaceae	*Trichostachys aurea* Hiern	Herb.		-X-
Rubiaceae	*Uncaria africana* G.Don		-X-	-X-
Rubiaceae	*Virectaria procumbens* (Sm.) Bremek.		Herb.	Herb.
Rubiaceae	*Virectaria* sp.		Herb.	
Rutaceae	*Vepris* sp.		Herb.	
Rutaceae	*Vepris verdoorniana* (Exell & Mendonça) W.Mziray	Herb.		
Rutaceae	***Zanthoxylum psammophilum* (Aké Assi) Waterman**		Herb.	
Rutaceae	*Zanthoxylum* sp.	-X-		
Sapindaceae	*Allophylus* sp.		Herb.	
Sapindaceae	*Chytranthus carneus* Radlk.	Herb.	Herb.	Herb.
Sapindaceae	*Chytranthus* sp.	Herb.		
Sapindaceae	*Eriocoelum racemosum* Baker		Herb.	

continued

Family	Species	North Lorma	Gola	Grebo
Sapindaceae	*Pancovia* sp.	Herb.		
Sapindaceae	*Paullinia pinnata* Linné	-X-		
Sapindaceae	***Placodiscus pseudostipularis* Radlk.**		Herb.	
Sapotaceae	*Chrysophyllum africanum* A.DC.	Herb.	Herb.	Herb.
Sapotaceae	*Chrysophyllum subnudum* Baker	Herb.		
Sapotaceae	*Chrysophyllum welwitschii* Engl.	Herb.		Herb.
Sapotaceae	*Delpydora gracilis* A.Chev.		Herb.	
Sapotaceae	*Englerophytum* sp.		Herb.	
Sapotaceae	*Gluema ivorensis* Aubrév. & Pellegr.		Herb.	
Sapotaceae	*Ituridendron bequaertii* De Wild.			Herb.
Sapotaceae	*Manilkara* sp.		Herb.	
Sapotaceae	*Neolemonniera* sp. Heine	Herb.		
Sapotaceae	*Pouteria aningeri* Baehni	Herb.		Herb.
Scytopetalaceae	*Scytopetalum tieghemii* Hutch. & Dalziel			Herb.
Selaginellaceae	*Selaginella cathedrifolia* Spring	Herb.	Herb.	
Selaginellaceae	*Selaginella myosurus* (Swartz) Alston			-X-
Selaginellaceae	*Selaginella soyauxii* Hieron.	Herb.		
Selaginellaceae	*Selaginella versicolor* Spring	Herb.		
Simaroubaceae	*Hannoa klaineana* Pierre ex Engl.	Herb.		
Solanaceae	*Solanum terminale* Forssk.	Herb.		
Sterculiaceae	***Cola buntingii* Baker f.**		Herb.	Photo
Sterculiaceae	***Cola caricifolia* (G.Don) K.Schum.**			-X-
Sterculiaceae	*Cola heterophylla* (P.Beauv.) Schott. & Endl.	Herb.		
Sterculiaceae	*Cola lateritia* K.Schum.			-X-
Sterculiaceae	*Cola* sp.	Herb.		
Sterculiaceae	***Heritiera utilis* Sprague**	Herb.	Herb.	-X-
Sterculiaceae	***Leptonychia occidentalis* Keay**			Herb.
Sterculiaceae	*Sterculia* sp.			Herb.
Sterculiaceae	*Triplochiton scleroxylon* K.Schum.			-X-
Thelypteridaceae	*Cyclosorus striatus* (Schum.) Ching	Herb.		
Thymelaeaceae	*Dicranolepis* sp.	-X-		-X-
Tiliaceae	*Desplatsia chrysochlamys* (Mildbr. & Burret) Mildbr. & Burret	Herb.		Herb.
Tiliaceae	*Grewia malacocarpa* Mast.	Herb.		
Tiliaceae	*Grewia pubescens* P.Beauv.	Herb.		
Urticaceae	*Urera* sp.	-X-		
Verbenaceae	***Vitex phaeotricha* Mildbr. ex W.Piep.**			Herb.
Violaceae	*Decorsella paradoxa* A.Chev.		Herb.	Herb.
Violaceae	*Rinorea brachypetala* (Turcz.) Kuntze	Herb.		
Violaceae	*Rinorea breviracemosa* Chipp	Herb.	Herb.	
Violaceae	*Rinorea ilicifolia* (Welw. ex Oliv.) Kuntze	Herb.		Herb.
Violaceae	***Rinorea microdon* M.Brandt**	Herb.	Herb.	
Violaceae	*Rinorea oblongifolia* (C.H.Wright) Marquand ex Chipp	Herb.		
Violaceae	*Rinorea* sp.			Herb.

continued

Family	Species	North Lorma	Gola	Grebo
Vitaceae	*Cissus diffusiflora* (Baker) Planch.	Herb.		
Vitaceae	***Cissus miegei*** **Tchoumé**			Herb.
Vitaceae	*Cissus producta* Afzel.			Herb.
Vitaceae	*Cissus smithiana* (Baker) Planch.			Herb.
Vitaceae	*Cissus* sp.	Herb.		
Vitaceae	*Leea guineensis* G.Don			-X-
Vittariaceae	*Antrophyum mannianum* Hook.	Herb.		
Vittariaceae	*Vittaria guineensis* Desv.	Herb.		
Zingiberaceae	*Aframomum* sp.		-X-	
Zingiberaceae	***Renealmia longifolia*** **K.Schum.**		Herb.	

Appendix 2

Checklist of Odonata recorded from Liberia and neighboring areas.

Klaas-Douwe B. Dijkstra

RL: Unpublished global or western African (between brackets) Red List assessment made by the author (assessed May, evaluated August 2006).

Biology (preferences are inferred from observations during the fieldwork, augmented with previous experience):
B: biogeography of the species. **A:** all over tropical Africa including savannahs, **G:** confined to Guineo-Congolian forest, **N:** associated with northern African savannah (Senegal to Ethiopia), **U:** confined to Upper Guinean forest (Sierra Leone to Togo), **W:** confined to western Africa forest (Senegal to Cameroon).
L: preferred landscape. **F:** forest, **O:** open habitats.
W: preferred water type. **R:** running; **S:** standing.

Liberian records (type locality lies in Liberia if species marked with asterisk):
NL, Go, Gr: North Lorma, Gola and Grebo National Forests.
A: adult voucher obtained; **L:** larval voucher obtained; **S:** adults caught for identification or seen only; records obtained nearby but outside the national forest are given between brackets.
Li: country records after Lempert (1988) and current survey. **1:** species found in current survey (! indicates new national record), **2:** found by Lempert, **3:** found by Lempert, but identification requires confirmation, **4:** literature record listed by Lempert; **5:** not listed by Lempert, but by Pinhey (1984). Species with old or dubious records (probable misidentifications) that are removed from the list until confirmed are: *Sapho orichalcea* McLachlan, 1869; *Umma puella* (Sjöstedt, 1917); *Ceriagrion ignitum* Campion, 1914; *Trithemis nuptialis* Karsch, 1894.

Neighboring areas (type locality lies in stated area if species marked with asterisk):
SL: Sierra Leone records after Carfi and D'Andrea (1994) and Marconi and Terzani (2006). **1:** authors' material; **2:** authors' material, identification requires confirmation; **3:** Aguesse (1968) records; **4:** other literature records. Omitted are: *Stenocnemis pachystigma* (Selys, 1886); *Elattoneura pruinosa* (Selys, 1886); *Agriocnemis forcipata* Le Roi, 1915; *Pseudagrion nubicum* Selys, 1876; *Anaciaeschna triangulifera* McLachlan, 1896; *Anax speratus* Hagen, 1867; *Diastatomma* sp. Gambles, 1987; *Phyllogomphus aethiops* Selys, 1854; *Phyllomacromia monoceros* (Förster, 1906); *Orthetrum caffrum* (Burmeister, 1839); *Orthetrum machadoi* Longfield, 1955; *Porpax asperipes* Karsch, 1896; *Trithemis dorsalis* (Rambur, 1842).
MN: Mt Nimba (Guinean side) records after Legrand (2003). **1:** author's material; **2:** author's material, identification requires confirmation; **3:** uncertain records, mostly personal communication P. Aguesse. Omitted are: *Lestes tridens* McLachlan, 1895; *Phyllomacromia aequatorialis* Martin, 1907; *Trithemis furva* Karsch, 1899.
Si: Simandou (Guinea) records after Legrand and Girard (1992). **1:** identification reliable; **2:** identification requires confirmation.
TF: Tai Forest (Côte d'Ivoire) records after Legrand and Couturier (1985): 1.

Taxa	Notes	RL	Biology			Liberian records				Neighboring areas			
			B	L	W	NL	Go	Gr	Li	SL	MN	Si	TF
Calopterygidae													
Phaon camerunensis Sjöstedt, 1900	1.		G	F	R	A		A	1	2	1	1	1
Phaon iridipennis (Burmeister, 1839)			A	O	R	A	A		1	1	1	1	1
Sapho bicolor Selys, 1853			G	F	R	A	A	S	1	1	1	1	1
Sapho ciliata (Fabricius, 1781)			W	F	R	A	A	A	1	1	1	1	1
Sapho fumosa Longfield, 1932	2.	NT	U	F	R		A		1	3*	1		
Umma cincta (Hagen in Selys, 1853)			G	F	R	A	A	S	1		1	1	1
Chlorocyphidae													
Chlorocypha curta (Hagen in Selys, 1853)			G	O	R				2	1	1	1	
Chlorocypha dispar (Palisot de Beauvois, 1807)			G	F	R	A	A	A	1	1	1	1	1
Chlorocypha luminosa (Karsch, 1893)	3.		U	F	R				2		1		
Chlorocypha pyriformosa Fraser, 1947	4.		G	F	R	A		S	1	1			1
Chlorocypha radix Longfield, 1959	5.		W	F	R	A	S	A	1	1	1	1	1
Chlorocypha rubida (Hagen in Selys, 1853)			W	F	R				2	1	3		1
Chlorocypha selysi (Karsch, 1899)			G	F	R	A	A	A	1	1	1	1	1
Lestidae													
Lestes dissimulans Fraser, 1955			A	O	S						1		1
Platycnemididae													
Mesocnemis singularis Karsch, 1891			A	O	R	S	S	A	1	1	1		1
Mesocnemis tisi Lempert, 1992	6.	EN	U	F	R				2*				
Platycnemis guttifera Fraser, 1950			W	F	R	A		A	1				1
Platycnemis sikassoensis (Martin, 1912)			G	O	R				2	1	1	1	1
Protoneuridae													
Chlorocnemis elongata Hagen in Selys, 1863			W	F	R	A	S	A	1	1	1	1	
Chlorocnemis flavipennis Selys, 1863	7.		W	F	R		A	A	1	1	1	1	
Chlorocnemis subnodalis (Selys, 1886)	8.		W	F	R	A	A	A	1	1	1		1
Elattoneura balli Kimmins, 1938			W	F	R	A	A	A	1	1*	1	1	1
Elattoneura dorsalis Kimmins, 1938		VU	U	F	R					1*			
Elattoneura girardi Legrand, 1980	9.		W	F	R				2	1	1		1
Elattoneura nigra Kimmins, 1938			G	O	R					1	1		
Prodasineura villiersi Fraser, 1948			U	F	R	A	A	A	1		1		1
Coenagrionidae													
Aciagrion africanum Martin, 1908			G	O	S				2		1		
Aciagrion gracile (Sjöstedt, 1909)			A	O	S						1		1
Africallagma subtile (Ris, 1921)	10.		A	O	S					1	1	1	
Agriocnemis angustirami Pinhey, 1974		VU	U	?	S				2*	1			
Agriocnemis exilis Selys, 1872			A	O	S				2	1			
Agriocnemis maclachlani Selys, 1877			G	F	S			A	1	3	1		1
Agriocnemis victoria Fraser, 1928	11.		G	O	S				2	1	3		
Agriocnemis zerafica Le Roi, 1915			A	O	S				2				
Argiagrion leoninum Selys, 1876	12.	DD	U	?	?					4*			
Ceriagrion bakeri Fraser, 1941			G	O	S			A	1	1	2		1

continued

Taxa	Notes	RL	Biology			Liberian records				Neighboring areas			
			B	L	W	NL	Go	Gr	Li	SL	MN	Si	TF
Ceriagrion corallinum Campion, 1914			G	O	S		A		1	1*			
Ceriagrion glabrum (Burmeister, 1839)			A	O	S	S	S		1	1	1	1	1
Ceriagrion rubellocerinum Fraser, 1947			G	F	S	A		A	1	1	1		1
Ceriagrion suave Ris, 1921	13.		A	O	S				3	1	2		
Ceriagrion tricrenaticeps Legrand, 1984		(DD)	G	?	S				2				
Ceriagrion whellani Longfield, 1952			A	O	S				2	3	1		
Ischnura senegalensis (Rambur, 1842)			A	O	S				2	1			
Pseudagrion aguessei Pinhey, 1964			N	O	R					3*			
Pseudagrion camerunense (Karsch, 1899)	14.		W	O	R				2	1			
Pseudagrion epiphonematicum Karsch, 1891			G	F	R	A	A	A	1	3	1	1	
Pseudagrion gigas Ris, 1936			N	?	R					3	1		
Pseudagrion glaucescens Selys, 1876			A	O	S				2	1	3		
Pseudagrion glaucoideum Schmidt in Ris, 1936			G	F	S	S			1				
Pseudagrion glaucum (Sjöstedt, 1900)	15.		G	O	S				2				
Pseudagrion hamoni Fraser, 1955			A	O	S					1		2	
Pseudagrion hemicolon Karsch, 1899	16.		G	F	R	A	A	A	1	1		1	1
Pseudagrion kersteni Gerstäcker, 1869			A	O	R					1			
Pseudagrion mascagnii Terzani & Marconi, 2004		CR	U	?	?					1*			
Pseudagrion melanicterum Selys, 1876			G	O	R	A	A	A	1	1	1	1	1
Pseudagrion sjoestedti Förster, 1906			A	O	R		A	S	1	1			1
Pseudagrion sublacteum (Karsch, 1893)			A	O	S				2	1		2	1
"*Pseudagrion*" *cyathiforme* Pinhey, 1973	17.		W	F	R				2	1			
"*Pseudagrion*" *malagasoides* Pinhey, 1973	18.		W	F	R				2				
Aeshnidae													
Anax chloromelas Ris, 1911			A	O	S					4			
Anax imperator Leach, 1815			A	O	S	S			1			1	
Anax tristis Hagen, 1867			A	O	S				2		3		
Gynacantha africana (Palisot de Beauvois, 1807)			G	F	S								1
Gynacantha bullata Karsch, 1891			G	F	S	S		A	1	1	1	1	1
Gynacantha cylindrata Karsch, 1891			G	F	S				2	1	3		1
Gynacantha manderica Grünberg, 1902			A	O	S					1	3		
Gynacantha nigeriensis (Gambles, 1956)	19.		G	F	S					1			
Gynacantha sextans McLachlan, 1896			G	F	S						1		1
Gynacantha sp. indet.	20.		?	?	?				3				
Gynacantha vesiculata Karsch, 1891			G	F	S				2	1	3		
Heliaeschna fuliginosa Karsch, 1893	21.		G	F	S		A	A	1	1			1
Heliaeschna cf. *cynthiae* Fraser, 1939	22.		?	?	?				3				
Gomphidae													
Diastatomma gamblesi Legrand, 1992	23.		U	F	R				2		1*		
Gomphidia bredoi (Schouteden, 1934)	24.		N	O	R								1
Gomphidia gamblesi Gauthier, 1987			W	F	R			S	1		1		
Ictinogomphus ferox (Rambur, 1842)			A	O	R					1	3		
Ictinogomphus fraseri Kimmins, 1958			W	F	R					1*			

continued

Taxa	Notes	RL	Biology			Liberian records				Neighboring areas			
			B	L	W	NL	Go	Gr	Li	SL	MN	Si	TF
Lestinogomphus africanus (Fraser, 1926)		DD	?	F	R					4*			
Lestinogomphus matilei Legrand & Lachaise, 2001	25.		U	F	R				2				
Lestinogomphus n. sp. 1	26.		U	F	R				3				
Lestinogomphus n. sp. 2	27.		U	F	R				3				
Lestinogomphus sp. indet.			?	F	R	S		L	1				
Microgomphus jannyae Legrand, 1992			U	F	R						1*		
Microgomphus sp. indet.	28.		?	F	R	A			1				
Onychogomphus xerophilus Fraser, 1956	29.	(DD)	U	F	R					2	2		
Paragomphus genei (Selys, 1841)			A	O	S				2	1			
Paragomphus kiautai Legrand, 1992		DD	U	F	R						1*		
Paragomphus mariannae Legrand, 1992	30.	DD	U	F	R				2		1*		
Paragomphus nigroviridis Cammaerts, 1968			G	F	R		A		1!				
Paragomphus serrulatus (Baumann, 1898)	31.		N	F	R				2	1			
Paragomphus tournieri Legrand, 1992	32.	DD	U	F	R				2		1*		
Paragomphus n. sp. cf. *elpidius* Ris, 1921	33.		U	F	R				2				
Paragomphus sp. indet.			?	F	R	L	L		1				
Phyllogomphus bartolozzii Marconi, Terzani & Carletti, 2001	34.	DD	U	F	R					1*			
Phyllogomphus helenae Lacroix, 1921		DD	U	F	R					4*			
Phyllogomphus moundi Fraser, 1960			W	F	R		A	S	1!		1		
Phyllogomphus n. sp.	35.		U	F	R				2				
Tragogomphus christinae Legrand, 1992	36.	DD	U	F	R				2		1*		
Tragogomphus sp. indet.			?	F	R		L		1				
Corduliidae													
Idomacromia lieftincki Legrand, 1984			G	F	R				2		1		
Idomacromia proavita Karsch, 1896			G	F	R						1		1
Neophya rutherfordi Selys, 1881			G	F	R				2	1	1		
Phyllomacromia aeneothorax (Nunney, 1895)		(DD)	G	F	R				2	4	1	1	
Phyllomacromia contumax Selys, 1879	37.		A	O	R				2				
Phyllomacromia funicularioides (Legrand, 1983)		NT	U	F	R				2		1*		
Phyllomacromia hervei (Legrand, 1980)			G	F	R	A		A	1				
Phyllomacromia kimminsi (Fraser, 1954)			A	F	R					1*			
Phyllomacromia lamottei (Legrand, 1993)	38.	DD	U	F	R				2		1*		
Phyllomacromia melania (Selys, 1871)	39.		G	F	R	A	A		1	4	1		
Phyllomacromia occidentalis (Fraser, 1954)		(DD)	U	F	R				2				
Phyllomacromia sophia (Selys, 1871)			U	F	R		A		1	4	1		1
Libellulidae													
Acisoma panorpoides Rambur, 1842			A	O	S	S	S	(S)	1	1	1		1
Acisoma trifidum Kirby, 1889			G	O	S	S	S	(S)	1	1	1		
Aethiothemis bella (Fisher, 1939)	40.	(DD)	G	F	?					3			
Aethiothemis solitaria Martin, 1908			A	O	S					2			
Aethriamanta rezia Kirby, 1889			A	O	S	S	S	S	1		1		
Atoconeura luxata Dijkstra, 2006	41.	(VU)	G	F	R					1	1		

continued

Taxa	Notes	RL	Biology			Liberian records				Neighboring areas			
			B	L	W	NL	Go	Gr	Li	SL	MN	Si	TF
Brachythemis lacustris (Kirby, 1889)			A	O	S				2				
Bradinopyga strachani (Kirby, 1900)			N	O	S		A		1	1	3		
Chalcostephia flavifrons Kirby, 1889			A	O	S	S	S	S	1	1	1		
Crocothemis divisa Baumann, 1898	42.		A	O	S				5	1	1		
Crocothemis erythraea (Brullé, 1832)			A	O	S	S	S	(S)	1	1	3		
Crocothemis sanguinolenta (Burmeister, 1839)			A	O	R		(S)		1	1	1	1	
Cyanothemis simpsoni Ris, 1915			G	F	R	A	S	S	1	4*	1		1
Diplacodes deminuta Lieftinck, 1969	43.	(DD)	A	O	S				2				
Diplacodes lefebvrii (Rambur, 1842)			A	O	S		A	(S)	1	1	1		
Diplacodes luminans (Karsch, 1893)	44.		A	O	S				2	1	3		
Eleuthemis buettikoferi Ris, 1910			G	F	R			S	1*	2	2		
Eleuthemis n. sp.			U	F	R		A		1				
Hadrothemis camarensis (Kirby, 1889)	45.		G	F	S	S	A	A	1	1	1		1
Hadrothemis coacta (Karsch, 1891)			G	F	S			A	1		3		1
Hadrothemis defecta (Karsch, 1891)			G	F	S	S	A	S	1	1	1		
Hadrothemis infesta (Karsch, 1891)			G	F	S	A	A	S	1	4	1		1
Hadrothemis versuta (Karsch, 1891)			G	F	S	S	A		1		1		
Hemistigma albipunctum (Rambur, 1842)			A	O	S		A		1	1	3		
Lokia incongruens (Karsch, 1893)	46.		W	F	R				2				
Malgassophlebia bispina Fraser, 1958			G	F	R				2		1		
Micromacromia camerunica Karsch, 1890	47.		G	F	R	A			1	4			
Micromacromia zygoptera (Ris, 1909)	48.		G	F	R			A	1	1	1		1
Neodythemis campioni Ris, 1915	49.	NT	U	F	R				2	4*	2		
Neodythemis klingi (Karsch, 1890)	50.		G	F	R		A	A	1	1	1		1
Nesciothemis minor Gambles, 1966			N	O	R			(A)	1!	1	1		
Nesciothemis nigeriensis Gambles, 1966			N	O	S					3			
Nesciothemis pujoli Pinhey, 1971	51.		N	O	S				4	1	2		
Notiothemis robertsi Fraser, 1944			G	F	S		S		1		1		
Olpogastra lugubris Karsch, 1895			A	O	R		A	S	1	3	1	1	
Orthetrum abbotti Calvert, 1892			A	O	S	A	A		1	1	1		
Orthetrum africanum (Selys, 1887)			G	F	R				2	4	1		
Orthetrum angustiventre (Rambur, 1842)			A	O	S				2	1			
Orthetrum austeni (Kirby, 1900)			G	O	S	S	S	S	1	1*	1		1
Orthetrum brachiale (Palisot de Beauvois, 1817)			A	O	S		A		1	1	3		1
Orthetrum chrysostigma (Burmeister, 1839)			A	O	S				2	1	3		
Orthetrum guineense Ris, 1910			A	O	R				4	1	1		
Orthetrum hintzi Schmidt, 1951			A	O	S		A	S	1	1	1		1
Orthetrum icteromelas Ris, 1910	52.		A	O	S				2	4			
Orthetrum julia Kirby, 1900			A	O	R	A	A	S	1	1*	1	1	1
Orthetrum latihami Pinhey, 1966	53.		N	O	?					1	1		
Orthetrum microstigma Ris, 1911			G	O	S	A	A	S	1	1	1	1	1
Orthetrum monardi Schmidt, 1951			A	O	S					1			
Orthetrum sagitta Ris, 1915		NT	U	?	?					2*			

continued

Taxa	Notes	RL	Biology			Liberian records				Neighboring areas			
			B	L	W	NL	Go	Gr	Li	SL	MN	Si	TF
Orthetrum stemmale (Burmeister, 1839)			A	O	S	A	A		1		3	1	1
Oxythemis phoenicosceles Ris, 1910			G	F	S				2				
Palpopleura deceptor (Calvert, 1899)			A	O	S		A		1!	4			
Palpopleura jucunda (Rambur, 1842)			A	O	S					1			
Palpopleura lucia (Drury, 1773)			A	O	S	S	S	S	1	1	1	1	1
Palpopleura portia (Drury, 1773)	54.		A	O	S	S	A	S	1	1	1	1	1
Pantala flavescens (Fabricius, 1798)			A	O	S	S	S	(S)	1	1	1		
Parazyxomma flavicans (Martin, 1908)			G	O	S	S			1				
Porpax bipunctus Pinhey, 1966		(VU)	G	F	?				2				1
Rhyothemis fenestrina (Rambur, 1842)			A	O	S	S	A	(S)	1	4			
Rhyothemis notata (Fabricius, 1781)			G	O	S			(A)	1	4	1		
Rhyothemis semihyalina (Desjardins, 1832)			A	O	S				2	1			
Sympetrum navasi Lacroix, 1921			A	O	S				2	3			
Tetrathemis camerunensis (Sjöstedt, 1900)	55.		G	F	S	A	A	S	1		3	1	1
Tetrathemis godiardi Lacroix, 1921			W	F	S		A	A	1		1		
Tetrathemis polleni (Selys, 1869)			A	O	S	S			1!				
Thermochoria equivocata Kirby, 1889			G	F	S				2	1	1		1
Tholymis tillarga (Fabricius, 1798)			A	O	S		A	A	1	1	1		1
Tramea basilaris (Palisot de Beauvois, 1817)			A	O	S	S	S	(S)	1	1			
Tramea limbata (Desjardins, 1832)			A	O	S	S	A	(S)	1!	1			
Trithemis aconita Lieftinck, 1969	56.		A	O	R	A	A	A	1	1	1	1	1
Trithemis africana (Brauer, 1867)		NT	U	F	R		A		1	1*			
Trithemis annulata (Palisot de Beauvois, 1807)			A	O	S				2	1	3		
Trithemis arteriosa (Burmeister, 1839)			A	O	S	A	A	S	1	1	1		1
Trithemis basitincta Ris, 1912			W	F	R		A		1				
Trithemis bredoi Fraser, 1953			N	O	S					1			
Trithemis dichroa Karsch, 1893			G	O	R				2	1	1		
Trithemis dejouxi Pinhey, 1978	57.		N	O	R				3				
Trithemis grouti Pinhey, 1961	58.		G	O	R	A	A	S	1	1	1	1	1
Trithemis hecate Ris, 1912	59.		A	O	?				3	1			
Trithemis kalula Kirby, 1900			N	O	?					1*	1		
Trithemis kirbyi Selys, 1891			A	O	S				2		3		
Trithemis monardi Ris, 1931	60.		A	O	S		A		1!	1			
Trithemis pruinata Karsch, 1899			G	F	R						1		
Trithemis stictica (Burmeister, 1839)			A	O	R				2	1			
Urothemis assignata (Selys, 1872)			A	O	S	S			1		1		
Urothemis edwardsii (Selys, 1849)			A	O	S				2	1			
Zygonyx chrysobaphes Ris, 1915			U	F	R			S	1	4*	1		
Zygonyx flavicosta (Sjöstedt, 1900)	61.		G	F	R	L	A		1	1	1		
Zygonyx geminunca Legrand, 1997	62.		U	F	R				3		1*		
Zygonyx torridus (Kirby, 1889)			A	O	R				2	1*	1		
Zyxomma atlanticum Selys, 1889			A	O	S		S	S	1				

Notes:

1. Includes Lempert's (1988) "*Phaon* cf. *fraseri* Pinhey, 1961";
2. The author's study of type specimens of *Sapho fumosa* and *Umma infumosa* Fraser, 1951 in the Natural History Museum in London suggest the two are synonymous;
3. Formerly known as *C. sharpae* Pinhey, 1972;
4. Formerly known as *C. mutans* Legrand & Couturier, 1986, misidentified as *C. neptunus* (Sjöstedt, 1899) by Carfi & D'Andrea (1994);
5. Formerly listed as *C. glauca radix* or just *C. glauca* (Selys, 1879);
6. "*Mesocnemis* sp.nov." in Lempert (1988);
7. *C. rossii* Pinhey, 1969, treated as a good species by Legrand (2003) is considered a synonym of *C. flavipennis* by the author;
8. Formerly placed in *Isomecocnemis*;
9. Misidentified as *E. acuta* Kimmins, 1938 by Carfi & D'Andrea (1994);
10. Formerly placed in *Enallagma*;
11. Misidentified as *A. forcipata* Le Roi, 1915 by Carfi & D'Andrea (1994);
12. Single female holotype from Sierra Leone is unlike any known African species and may pertain to a mislabelling.
13. Includes *C. moorei* Longfield, 1952;
14. Formerly known as *P. angelicum* Fraser, 1947;
15. Formerly known as *P. basicornu* Schmidt in Ris, 1936;
16. Formerly known as *P. flavipes* Sjöstedt, 1899 or *P. f. leonense* Pinhey, 1964;
17. Described as *Aciagrion walteri* by Carfi & D'Andrea (1994);
18. This and the previous species probably belong to an unnamed genus;
19. Formerly known as *G. sevastopuloi* (Pinhey, 1961), identification by Carfi & D'Andrea (1994) confirmed;
20. Lempert's (1988) "*Gynacantha* sp." female could not be assigned to a known species;
21. Misidentified as "*Gynacantha* cfr. *usambarica* Sjöstedt, 1909" by Carfi & D'Andrea (1994);
22. Female published by Lempert (1988) as *H. fuliginosa*;
23. "*Diastatomma* sp. nov." in Lempert (1988);
24. Formerly known as *G. madi* Pinhey, 1961:
25. "*Lestinogomphus* sp. 2" in Lempert (1988);
26. "*Lestinogomphus* sp. 1" in Lempert (1988);
27. "*Lestinogomphus* sp. 3" in Lempert (1988);
28. Includes Lempert's (1988) "*Microgomphus* sp." females;
29. Records of *O. quirkii* Pinhey, 1964 and *O. supinus* Hagen in Selys, 1854 listed under this name, the only one for this type of *Onychogomphus* from West Africa;
30. "*Paragomphus* sp. nov. 3" in Lempert (1988);
31. Formerly known as *P. bredoi* (Schouteden, 1934) includes records by that name and "*Paragomphus* sp. nov. 2" by Lempert (1988) and misidentified as *P. cognatus* (Rambur, 1842) by Carfi & D'Andrea (1994);
32. "*Paragomphus* sp. nov. 4" in Lempert (1988);
33. "*Paragomphus* sp. nov. 1" in Lempert (1988);
34. May be the same as *P. moundi* and *P. helenae*;
35. Lempert's (1988) "*Phyllogomphus* sp." male is unlike known species;
36. Identified as *T. tenaculatus* (Fraser, 1926) by Lempert (1988);
37. Formerly known as *P. bifasciata* Martin, 1912;
38. Lempert's (1988) "*Macromia* sp. nov." is this species;
39. Formerly known as *P. funicularia* (Martin, 1907);
40. Formerly known as *Sleuthemis diplacoides* Fraser, 1951 and *Monardithemis leonensis* Aguesse, 1968;
41. Formerly mistaken for *A. biordinata* Karsch, 1899;
42. Misidentified as *C. saxicolor* Ris, 1921 by Carfi & D'Andrea (1994), old Liberian record of that species also included here;
43. Three specimens published by Lempert (1988) as *D. lefebvrii*;
44. Formerly placed in *Philonomon*;
45. Misidentified (partly) as *Lokia incongruens* by Carfi & D'Andrea (1994);
46. Several records accidentally excluded by Lempert (1988);
47. Some records may requires reexamination following confusion with *M. zygoptera*;
48. Formerly placed in *Eothemis*;
49. Formerly placed in *Allorrhizucha*, probably misidentified as *Neodythemis scalarum* Pinhey, 1964 by Legrand (2003);
50. Formerly placed in *Allorrhizucha*;
51. All western African specimens of *N. farinosa* (Förster, 1898) examined by author pertained to *N. pujoli*;
52. Single Liberian female among material not noted previously by Lempert (1988);
53. Identification by Carfi & D'Andrea (1994) confirmed;
54. *P. lucia* and *P. portia* were not separated by Lempert (1988), but both present in material;
55. Includes *T. bifida* Fraser, 1941;
56. Misidentified as *T. bifida* Pinhey, 1970 and *T. basitincta* by Carfi & D'Andrea (1994);
57. "*T. donaldsoni* (Calvert, 1899)" in Lempert (1988);
58. Formerly known as *T. atra* Pinhey, 1961, misidentified as *T. nuptialis* Karsch, 1894 by Carfi & D'Andrea (1994);
59. Identification by Marconi & Terzani (2006) confirmed, Lempert's (1988) "*T.* cf. *hecate*" is probably also correct;
60. Includes *T. imitata* Pinhey, 1961;
61. Includes *Z. fallax* (Schouteden, 1934);
62. Lempert's (1988) "*Zygonyx* sp." may be this species, but the specimen is lost.

Appendix 3

Locality list and short description of habitats investigated in North Lorma National Forest/ Wologizi (WOL), Gola National Forest (GO) and Grebo National Forest (GRE).

Annika Hillers and Mark-Oliver Rödel

Site	Latitude (N)	Longitude (W)	Date	Description
WOL1	8°01.741'	9°44.119'	20.11.2005	Primary forest with small stream
WOL2	8°01.929'	9°44.161'	20.11.2005	Dry primary forest on hill above big river
WOL3	8°01.434'	9°44.414'	21.11.2005	Primary forest, one part slightly swampy area, other part brook, next to big river
WOL4	8°01.523'	9°44.226'	21.11.2005	Dry forest over river, many lianas, thick undergrowth
WOL5	8°02.043'	9°43.970'	22.11.2005	Forest around stream with rocks and stones, further in forest sandy, and slightly swampy area with temporary puddles
WOL6	8°02.023'	9°44.143'	22.11.2005	Forest over river, on one side stream with small waterfall, many shrubs and bushs
WOL7	8°02.509'	9°43.682'	23.11.2005	Dry forest with big rocks and stones
WOL8	8°02.391'	9°43.750'	23.11.2005	Swampy area in forest, with Raffia and Marantaceae, partly open canopy
WOL9	8°01.722'	9°44.124'	24.11.2005	Forest with streams and partly swampy area
GO1	7°27.178'	10°41.522'	28.11.2005 & 1.12.2005	Hilly primary forest with stream, stream with rocks and sand
GO2	7°27.272'	10°41.548'	29.11.2005	Dry forest on hill, with some rocks
GO3	7°27.376'	10°41.736'	29.11.2005 & 1.12.2005	Old diamond mines and ponds within forest, partly open area, and forest around this area
GO4	7°27.293'	10°41.632'	29.11.2005	Dry forest on hill
GO5	7°27.352'	10°41.483'	30.11.2005	Valley within forest with small brook, partly swampy area and forest around, partly on hill
GO6	7°26.781'	10°39.063'	2.12.2005	Big pond near SLC village
GO7	7°26.404'	10°39.150'	2.12.2005	Small pond next to big river Mano, with stones and a few trees
GRE1	5°24.108'	7°44.011'	7.12.2005	Mature secondary forest, partly thick undergrowth, with sandy stream and temporary puddles
GRE2	5°24.358'	7°44.106'	8.12.2005 & 10.12.2005	Swampy area in forest with small stream, with many treefall gaps and lianas, forest around the swampy area
GRE3	5°24.535'	7°44.276'	8.12.2005	Dry forest
GRE4	5°24.285'	7°43.786'	9.12.2005	Swampy area within secondary forest near stream with many lianas and shrubs, thick leaf litter coverage
GRE5	5°24.334'	7°43.631'	9.12.2005	Dry forest on hill
GRE6	5°23.857'	7°42.536'	10.12.2005	Big pond next to old logging road in mature secondary forest
GRE7	5°24.083'	7°42.892'	10.12.2005	Pond next to old logging road in mature secondary forest
GRE8	5°24.286'	7°42.954'	10.12.2005	Small pond next to old logging road in mature secondary forest
GRE9	5°24.127' & 5°23.827'	7°43.965' & 7°44.160'	9.12.2005 & 11.12.2005	On or next to old logging road in mature secondary forest

Appendix 4

Amphibian species recorded in North Lorma, Gola and Grebo National Forests.

Annika Hillers and Mark-Oliver Rödel

Amphibian species recorded in North Lorma (WOL), Gola (GO) and Grebo (GRE) National Forests with record sites (see Appendix 3), habitat preference and their distribution in Africa.
S = savannah
FB = farmbush (degraded forest and farmland)
F = forest
A = Africa (occurs also outside West Africa)
WA = West Africa (Senegal to eastern Nigeria)
UG = Upper Guinea (forest zone West of the Dahomey Gap)
E = endemic to Liberia
* = records possibly comprise several species
** = CITES listed species
spp. & cf. = determination needs confirmation or new species are involved
LC = Least concern
NT = Near Threatened
VU = Vulnerable
EN = Endangered
[1] = first country record

Frost et al. (2006) introduced many new names and relationships. As these are not yet generally accepted and to allow for a better orientation with older literature, we herein list the old names. The new affiliations according to Frost et al. (2006) are: The West African *Bufo* species are now in the genus *Amietophrynus.* The African *Amnirana* species in the genus *Hydrophylax. Astylosternus* and *Leptopelis* moved into the family Arthroleptidae. *Conraua* now belongs into the family Petropedetidae, *Hoplobatrachus* into the family Dicroglossidae, *Ptychadena* into the family Ptychadenidae and *Phrynobatrachus* forms the family Phrynobatrachidae.

Species	Site	S	FB	F	A	WA	UG	E	IUCN Red List category
Arthroleptidae									
Arthroleptis spp. *	WOL 1,6,7 GO 1,5 GRE 1,2,4,5,9		x	x			x		LC
Cardioglossa leucomystax	WOL 1,5 GO 1,5 GRE 1			x	x				LC
Astylosternidae									
Astylosternus occidentalis[1]	GO 1,5			x			x		LC
Bufonidae									
Bufo maculatus	WOL 6 GO 3,6 GRE 7,9	x	x	x	x				LC

continued

Species	Site	S	FB	F	A	WA	UG	E	IUCN Red List category
Bufo regularis	WOL 6 GO 6 GRE 9	x	x		x				LC
*Bufo superciliaris***[1]	WOL 1,3			x	x				LC
Bufo togoensis	WOL 1 GRE 1			x			x		NT
Hyperoliidae									
Afrixalus dorsalis	GO 3,6 GRE 8	x	x	x	x				LC
Afrixalus nigeriensis[1]	GO 3 GRE 1,7,8		x	x		x			NT
Hyperolius chlorosteus	GO 1 GRE 1,4		x	x			x		NT
Hyperolius concolor	GO 6	x	x	x		x			LC
Hyperolius fusciventris	GO 3,6 GRE 6,7		x	x		x			LC
Hyperolius guttulatus	GO 3,6 GRE 6,7		x	x	x				LC
Hyperolius picturatus	GO 6 GRE 1,2		x	x			x		LC
Leptopelis hyloides	WOL 1,5 GO 6 GRE 1		x	x		x			LC
Leptopelis macrotis	GRE 1			x			x		NT
Leptopelis occidentalis	GRE 1			x			x		NT
Phlyctimantis boulengeri	GRE 8		x	x		x			LC
Petropedetidae									
Petropedetes natator	GO 1			x			x		NT
Phrynobatrachus accraensis	GO 6	x	x			x			LC
Phrynobatrachus alleni	WOL 1,2,3,5,6,8,9 GO 4 GRE 4			x			x		NT
Phrynobatrachus annulatus	WOL 3,6			x			x		EN
Phrynobatrachus cf. *annulatus*	GRE 2			x			(x)		(EN)
Phrynobatrachus fraterculus	WOL 1,3,5,9 GO 3			x			x		LC
Phrynobatrachus guineensis	GRE 2,4			x			x		NT
Phrynobatrachus liberiensis	WOL 1,2,3,5 GO 1,5 GRE 1,2,4			x			x		NT
Phrynobatrachus phyllophilus	WOL 1,5 GO 1,3,5 GRE 2,4			x			x		NT
Phrynobatrachus plicatus	WOL 1,6,9 GO 1 GRE 2,4,9			x		x			LC

continued

Species	Site	S	FB	F	A	WA	UG	E	IUCN Red List category
Phrynobatrachus tokba	WOL 1,3 GO 1,4,5 GRE 2,9			x			x		LC
Phrynobatrachus villiersi[1]	WOL 2 GO 5 GRE 2			x			x		VU
Pipidae									
Silurana tropicalis	GRE 1		x	x	x				LC
Rhacophoridae									
Chiromantis rufescens[1]	WOL 6 GO 3, 7 GRE 8			x	x				LC
Ranidae									
Amnirana albolabris	WOL 1 GO 6 GRE 1,2,4		x	x	x				LC
Amnirana occidentalis	GO 1			x			x		EN
Conraua alleni	GO 1 GRE 2			x			x		VU
Hoplobatrachus occipitalis	GO 3,6	x	x	x	x				LC
Ptychadena aequiplicata	WOL 1,4,5 GO 3 GRE 2			x	x				LC
Ptychadena bibroni	GO 3,6	x	x		x				LC
Ptychadena longirostris	GO 3 GRE 9		x	x		x			LC
Ptychadena superciliaris	GRE 6			x			x		NT

Appendix 5

Amphibian tissue samples and voucher specimens collected during Liberia RAP survey.

Annika Hillers and Mark-Oliver Rödel

List of tissue samples (DNA) and voucher specimens.
MOR = collection Rödel; tissue samples preserved in 95% ethanol, voucher specimens preserved in 70% ethanol.

North Lorma (WOL) National Forest
Gola (GO) National Forest
Grebo (GRE) National Forest

Species	DNA	MOR
Arthroleptis spp.	WOL 15, 16, 22, 56, 68, 74, 83, 84, 86, 94 GO 6, 21, 22, 23, 45, 55, 72 GRE 1, 2, 3, 7, 8, 11, 15, 16, 18, 20, 21, 22, 23, 24, 25, 26, 27, 31, 55, 58, 76, 78, 79, 80, 81	WOL 15, 68, 74, 94 GO 6, 45 GRE 11, 58
Cardioglossa leucomystax	WOL 87, 88, 89, 98 GO 7, 43, 50 GRE 56	WOL 87 GO 43 GRE 56
Astylosternus occidentalis	GO 4, 29, 58	GO 4
Bufo regularis	GRE 73, 74	GRE73
Bufo superciliaris	WOL 19, 78, 79	
Bufo togoensis	WOL 80, 18, 23, 80, GRE 28, 29, 36	WOL 80 GRE36
Afrixalus dorsalis	GO 60	GO 60
Afrixalus nigeriensis	GO 61 GRE 43	GO 61 GRE 43
Hyperolius chlorosteus	GRE 5, 46, 64	GRE 46
Hyperolius fusciventris	GO 16, 17, 62, 64, 65, 67	GO 62, 64, 65, 67
Hyperolius guttulatus	GO 63, 66 GRE 72	GO 63 GRE 72
Hyperolius picturatus	GRE 41	GRE 41
Leptopelis hyloides	WOL 101, 102, 103 GRE45	WOL 101 GRE 45
Leptopelis macrotis	GRE 40	GRE 40
Leptopelis occidentalis	GRE 42	GRE 42
Phlyctimantis boulengeri	GRE 75	GRE 75
Petropedetes natator	GO 15, 19, 39	GO 39

continued

Species	DNA	MOR
Phrynobatrachus alleni	WOL 1, 2, 3, 4, 6, 8, 10, 13, 20, 24, 25, 28, 34, 35, 38, 41, 43, 54, 59, 60, 61, 62, 67, 72, 73 GO 33, 35, 36 GRE 37, 59, 61, 67	WOL 13, 72, 73 GRE 37, 59
Phrynobatrachus annulatus	WOL 69, 82, 92	WOL 92
Phrynobatrachus cf. *annulatus*	GRE 47	GRE 47
Phrynobatrachus fraterculus	WOL 44, 46, 51, 52, 63, 97, 99, 100 GO 31, 37, 38, 44, 47	WOL 51, 52 GO 44
Phrynobatrachus guineensis	GRE 48, 62	GRE 48, 62
Phrynobatrachus liberiensis	WOL 5, 11, 21, 26, 27, 31, 32, 33, 34, 36, 37, 39, 40, 42, 50, 58, 81, 85 GO 27, 28, 41 GRE 6, 12, 13, 14, 17, 33, 53	WOL 50 GO 41 GRE 13, 53
Phrynobatrachus phyllophilus	WOL 7, 9, 12, 14, 17, 55, 64, 65, 93, 96 GO 5, 8, 25, 26, 30, 46, 48, 52 GRE 34, 60, 68	WOL 14, 93 GO 5, 46, 52 GRE 34
Phrynobatrachus plicatus	WOL 71, 76, 77, 90, 95 GO 42, 71 GRE 19, 30, 35, 49, 50, 51, 54	WOL 71 GO 42 GRE 30
Phrynobatrachus tokba	WOL 29, 30, 45, 53, 57 GO 2, 3, 9, 24, 32, 49, 51, 54 GRE 63, 65, 66, 69, 77, 82, 83, 84	WOL 53 GO 9 GRE 63
Phrynobatrachus villiersi	GRE 32	GRE 32
Chiromantis rufescens	GO 18, 53, 59 GRE 70	GO 53, 59 GRE 70
Amnirana albolabris	WOL 70 GO 73 GRE 4, 10	WOL 70 GO 73 GRE 10
Amnirana occidentalis	GO 10, 11, 12, 13, 14, 56, 57	GO 12, 14
Conraua alleni	GO 20, 40 GRE 38, 39, 44	GO 40 GRE 38, 39
Ptychadena aequiplicata	WOL 48, 49, 66, 75, 91 GRE 9	WOL 75 GRE 9
Ptychadena longirostris	GRE 52, 57	GRE 57
Ptychadena superciliaris	GRE 71	GRE 71

Appendix 6

Reptile species recorded in North Lorma, Gola and Grebo National Forests.

Annika Hillers and Mark-Oliver Rödel

Taxa	Site	CITES Appendix #
REPTILIA - SAURIA		
Agamidae		
Agama agama	Gola, Grebo	
Gekkonidae		
Hemidactylus aff. *muriceus*	Gola	
Scincidae		
Trachylepis affinis	Grebo	
Cophoscincopus sp.1	North Lorma, Gola	
Cophoscincopus sp. 2	Gola	
Varanidae		
Varanus ornatus	Gola	2
REPTILIA - SERPENTES		
Boidae		
Python sebae	North Lorma	2
Colubridae		
Dipsadoboa sp.	Grebo	
Grayia smythii	Gola	
Natriciteres variegata	Grebo	
Philothamnus heterodermus	North Lorma	
Rhamnophis aethiopissa	North Lorma	
Viperidae		
Bitis gabonica	Gola	
Atheris chlorechis	North Lorma, Gola	
REPTILIA - CHELONIA		
Testudinidae		
Kinixys homeana	North Lorma	2
Kinixys erosa	Grebo	2
REPTILIA - CROCODYLIA		
Crocodylidae		
Osteolaemus tetraspis	Gola, Grebo	1

Appendix 7

Bird species recorded in North Lorma, Gola and Grebo National Forests.

Ron Demey

Encounter rate :
C = Common: encountered daily, either singly or in significant numbers
F = Fairly common: encountered on most days
U = Uncommon: irregularly encountered and not on the majority of days
R = Rare: rarely encountered, one or two records of single individuals

Breeding :
b = evidence of breeding observed (nest with eggs or young, or juveniles with parents)

Threat Status :
EN = Endangered
VU = Vulnerable
DD = Data Deficient
NT = Near Threatened

Endemism :
UG = endemic to the Upper Guinea forest block

Biome :
GC = restricted to Guinea-Congo Forests biome

Habitat :
f = primary or old secondary forest
d = degraded or heavily logged forest
e = forest edge
o = open areas (large clearings, cultivation, etc)
w = rivers, streams, swamps and ponds
a = aerial and flying overhead

Taxa	Common Name	North Lorma	Gola	Grebo	Threat Status	Endemism	GC Forests Biome	Habitat
ARDEIDAE								
Tigriornis leucolopha	White-crested Tiger Heron			R			GC	w
Ixobrychus sturmii	Dwarf Bittern		R					w
Bubulcus ibis	Cattle Egret		R					o
Ardea cinerea / melanocephala	Grey / Black-headed Heron			R				a
CICONIIDAE								
Ciconia episcopus	Woolly-necked Stork		R					a
THRESKIORNITHIDAE								
Bostrychia hagedash	Hadada Ibis	U	R					w
Bostrychia olivacea	Olive Ibis	U						w
Bostrychia rara	Spot-breasted Ibis			R			GC	a
ANATIDAE								
Pteronetta hartlaubii	Hartlaub's Duck			R			GC	w

continued

Taxa	Common Name	North Lorma	Gola	Grebo	Threat Status	Endemism	GC Forests Biome	Habitat
ACCIPITRIDAE								
Gypohierax angolensis	Palm-nut Vulture	U	F	C				f, d, a
Dryotriorchis spectabilis	Congo Serpent Eagle			R			GC	d
Polyboroides typus	African Harrier Hawk	U	R	F				f, d, a
Accipiter tachiro	African Goshawk	U	U	F				f, d
Accipiter melanoleucus	Black Sparrowhawk		R					o
Urotriorchis macrourus	Long-tailed Hawk	R					GC	f
Stephanoaetus coronatus	Crowned Eagle			R				d
PHASIANIDAE								
Francolinus lathami	Latham's Forest Francolin	F	U	C			GC	f, d
Francolinus ahantensis	Ahanta Francolin		F	F			GC	e
NUMIDIDAE								
Agelastes meleagrides	White-breasted Guineafowl			U	VU	UG	GC	f
Guttera pucherani	Crested Guineafowl			R				d
RALLIDAE								
Himantornis haematopus	Nkulengu Rail			F			GC	d
Sarothrura pulchra	White-spotted Flufftail	F	F	C			GC	w
HELIORNITHIDAE								
Podica senegalensis	African Finfoot	U						w
GLAREOLIDAE								
Glareola nuchalis	Rock Pratincole		U					w
CHARADRIIDAE								
Vanellus albiceps	White-headed Lapwing	R						w
SCOLOPACIDAE								
Actitis hypoleucos	Common Sandpiper	U						w
COLUMBIDAE								
Treron calvus	African Green Pigeon	C	C	C				f, d, e
Turtur brehmeri	Blue-headed Wood Dove	C	C	C			GC	f, d
Turtur tympanistria	Tambourine Dove	F	F	U				d, e
Turtur afer	Blue-spotted Wood Dove	F	R	U				e, o
Columba iriditorques	Western Bronze-naped Pigeon	F	C	F			GC	f, d
Columba unicincta	Afep Pigeon	R					GC	f
Streptopelia semitorquata	Red-eyed Dove	F		F				e, o
PSITTACIDAE								
Psittacus erithacus	Grey Parrot	F	U	C			GC	f, d, a
Poicephalus gulielmi	Red-fronted Parrot			U				a
Agapornis swindernianus	Black-collared Lovebird			R			GC	d
MUSOPHAGIDAE								
Corythaeola cristata	Great Blue Turaco	C	R	F				f, d
Tauraco macrorhynchus	Yellow-billed Turaco	C	F	C			GC	f, d

continued

Taxa	Common Name	North Lorma	Gola	Grebo	Threat Status	Endemism	GC Forests Biome	Habitat
CUCULIDAE								
Oxylophus levaillantii	Levaillant's Cuckoo		R	R				e
Cuculus clamosus	Black Cuckoo			R				d
Cercococcyx mechowi	Dusky Long-tailed Cuckoo			F			GC	d
Cercococcyx olivinus	Olive Long-tailed Cuckoo	U	U	U			GC	f, d
Chrysococcyx cupreus	African Emerald Cuckoo	U	F	F				f, d
Chrysococcyx klaas	Klaas's Cuckoo	U	U					o
Chrysococcyx caprius	Didric Cuckoo		R					o
Ceuthmochares aereus	Yellowbill	C	F	F				f, d
Centropus leucogaster	Black-throated Coucal	U	U	F			GC	f, d
Centropus senegalensis	Senegal Coucal	R	R	R				e, o
STRIGIDAE								
Glaucidium tephronotum	Red-chested Owlet			R			GC	d
APODIDAE								
Rhaphidura sabini	Sabine's Spinetail	U	F	F			GC	a
Apus apus	Common Swift	F	C	C				a
TROGONIDAE								
Apaloderma narina	Narina's Trogon	F	U	F				f, d
ALCEDINIDAE								
Halcyon badia	Chocolate-backed Kingfisher	C	F	C			GC	f, d
Halcyon leucocephala	Grey-headed Kingfisher		R					o
Halcyon malimbica	Blue-breasted Kingfisher	F	F	U				f, d
Halcyon senegalensis	Woodland Kingfisher		R	R				o
Ceyx lecontei	African Dwarf Kingfisher	R	R	R			GC	f, d, e
Ceyx pictus	African Pygmy Kingfisher	R	U					o
Alcedo leucogaster	White-bellied Kingfisher	R		R			GC	w
Alcedo quadribrachys	Shining-blue Kingfisher	R	R	R				w
Megaceryle maxima	Giant Kingfisher	R						w
MEROPIDAE								
Merops muelleri	Blue-headed Bee-eater			R			GC	e
Merops gularis	Black Bee-eater		U	R			GC	e
Merops albicollis	White-throated Bee-eater	F	C	C				o, a
CORACIIDAE								
Eurystomus gularis	Blue-throated Roller	R	R				GC	f, d
Eurystomus glaucurus	Broad-billed Roller			U				o
PHOENICULIDAE								
Phoeniculus castaneiceps	Forest Wood-hoopoe	R		R			GC	f, d
BUCEROTIDAE								
Tropicranus albocristatus	White-crested Hornbill	U	U	R			GC	f, d
Tockus hartlaubi	Black Dwarf Hornbill			R			GC	e
Tockus camurus	Red-billed Dwarf Hornbill	C	F	U			GC	f, d

continued

Taxa	Common Name	North Lorma	Gola	Grebo	Threat Status	Endemism	GC Forests Biome	Habitat
Tockus fasciatus	African Pied Hornbill	F	F	F			GC	f, d, e
Bycanistes fistulator	Piping Hornbill	R	U	F			GC	f, d, e
Bycanistes cylindricus	Brown-cheeked Hornbill	C	C	C	NT	UG	GC	f, d
Ceratogymna atrata	Black-casqued Hornbill	F	C	C			GC	f, d
Ceratogymna elata	Yellow-casqued Hornbill	C	C	C	NT		GC	f, d
CAPITONIDAE								
Gymnobucco calvus	Naked-faced Barbet	C	F	C			GC	f, d, e
Pogoniulus scolopaceus	Speckled Tinkerbird	C	C	C			GC	f, d, e
Pogoniulus atroflavus	Red-rumped Tinkerbird	C	C	C			GC	f, d
Pogoniulus subsulphureus	Yellow-throated Tinkerbird	C	C	C			GC	f, d, e
Buccanodon duchaillui	Yellow-spotted Barbet	C	C	C			GC	f, d
Tricholaema hirsuta	Hairy-breasted Barbet	R	U	C			GC	f, d
Trachylaemus purpuratus	Yellow-billed Barbet			F			GC	d
INDICATORIDAE								
Prodotiscus insignis	Cassin's Honeybird		R	R			GC	d
Melignomon eisentrauti	Yellow-footed Honeyguide	R			DD		GC	f
Melichneutes robustus	Lyre-tailed Honeyguide	R					GC	f
Indicator maculatus	Spotted Honeyguide	R					GC	f
Indicator conirostris	Thick-billed Honeyguide		R	R				d, e
PICIDAE								
Campethera maculosa	Little Green Woodpecker	R		F			GC	d
Campethera nivosa	Buff-spotted Woodpecker	R		U			GC	f, d
Campethera caroli	Brown-eared Woodpecker	U		R			GC	f, d
Dendropicos gabonensis	Gabon Woodpecker	R	R	R			GC	d
Dendropicos pyrrhogaster	Fire-bellied Woodpecker	R	R	R			GC	d
EURYLAIMIDAE								
Smithornis rufolateralis	Rufous-sided Broadbill	F		F			GC	f, d
HIRUNDINIDAE								
Psalidoprocne nitens	Square-tailed Saw-wing	F	C	F			GC	o
Hirundo abyssinica	Lesser Striped Swallow		R					o
Hirundo nigrita	White-throated Blue Swallow		R				GC	w
Hirundo rustica	Barn Swallow	U	R					o
MOTACILLIDAE								
Motacilla flava	Yellow Wagtail		U					o
Motacilla clara	Mountain Wagtail	R	F					w
Motacilla aguimp	African Pied Wagtail		U					o, w
CAMPEPHAGIDAE								
Campephaga quiscalina	Purple-throated Cuckoo-shrike	R		R				f, d
Lobotos lobatus	Western Wattled Cuckoo-shrike			R	VU	UG	GC	d
Coracina azurea	Blue Cuckoo-shrike	C	C	C			GC	f, d

continued

Taxa	Common Name	North Lorma	Gola	Grebo	Threat Status	Endemism	GC Forests Biome	Habitat
PYCNONOTIDAE								
Andropadus virens	Little Greenbul	C	C	C				d, e
Andropadus gracilis	Little Grey Greenbul	U	U	F			GC	d, e
Andropadus ansorgei	Ansorge's Greenbul	C	C	C			GC	f, d, e
Andropadus curvirostris	Cameroon Sombre Greenbul	R	U	F			GC	f, d, e
Andropadus gracilirostris	Slender-billed Greenbul	C	C	C				f, d, e
Andropadus latirostris	Yellow-whiskered Greenbul	C	C	C				f, d
Calyptocichla serina	Golden Greenbul	R	F	U			GC	f, d, e
Baeopogon indicator	Honeyguide Greenbul	C	C	C			GC	f, d, e
Ixonotus guttatus	Spotted Greenbul	C	C	C			GC	f, d
Chlorocichla simplex	Simple Leaflove	R	C	U			GC	o
Thescelocichla leucopleura	Swamp Palm Bulbul	C	C	F			GC	f, d
Phyllastrephus icterinus	Icterine Greenbul	C	U	C			GC	f, d
Bleda syndactylus	Red-tailed Bristlebill	U	F	F			GC	f, d
Bleda eximius	Green-tailed Bristlebill			U	VU	UG	GC	d
Bleda canicapillus	Grey-headed Bristlebill	C		C			GC	f, d
Criniger barbatus	Western Bearded Greenbul	C	F	C			GC	f, d
Criniger calurus	Red-tailed Greenbul	C	F	C			GC	f, d
Criniger olivaceus	Yellow-bearded Greenbul	F	U	U	VU	UG	GC	f, d
Pycnonotus barbatus	Common Bulbul		F					o
Nicator chloris	Western Nicator	C	C	C			GC	f, d, e
TURDIDAE								
Stiphrornis erythrothorax	Forest Robin	F	U	C / b			GC	f, d
Luscinia megarhynchos	Common Nightingale		U					o
Cossypha cyanocampter	Blue-shouldered Robin Chat			R			GC	e
Alethe diademata	Fire-crested Alethe	C	F	C			GC	f, d
Alethe poliocephala	Brown-chested Alethe	R						f
Neocossyphus poensis	White-tailed Ant Thrush	C	F	C			GC	f, d
Stizorhina finschi	Finsch's Flycatcher Thrush	C	C	C			GC	f, d
Cercotrichas leucosticta	Forest Scrub Robin	U		R			GC	f, d
Turdus pelios	African Thrush		U					o
SYLVIIDAE								
Bathmocercus cerviniventris	Black-headed Rufous Warbler	R			NT	UG	GC	e
Hippolais pallida	Olivaceous Warbler		U					o
Cisticola lateralis	Whistling Cisticola	R						o
Cisticola brachypterus	Short-winged Cisticola		F					o
Prinia subflava	Tawny-flanked Prinia		F	U				o
Apalis nigriceps	Black-capped Apalis	R		C			GC	f, d
Apalis sharpii	Sharpe's Apalis	C	C / b	C		UG	GC	f, d
Camaroptera brachyura	Grey-backed Camaroptera	U	F					e, o

continued

Taxa	Common Name	North Lorma	Gola	Grebo	Threat Status	Endemism	GC Forests Biome	Habitat
Camaroptera superciliaris	Yellow-browed Camaroptera	C	C	F			GC	e, o
Camaroptera chloronota	Olive-green Camaroptera	C	C	C			GC	f, d, e
Macrosphenus kempi	Kemp's Longbill	U	C	F			GC	e
Macrosphenus concolor	Grey Longbill	C	C	C			GC	f, d, e
Eremomela badiceps	Rufous-crowned Eremomela		U	U			GC	d, e
Sylvietta virens	Green Crombec	F		U			GC	e
Sylvietta denti	Lemon-bellied Crombec	U		U			GC	d, e
Phylloscopus trochilus	Willow Warbler		U					o
Hyliota violacea	Violet-backed Hyliota		U	U			GC	d, e
Hylia prasina	Green Hylia	C	C	C / b			GC	f, d, e
MUSCICAPIDAE								
Fraseria ocreata	Fraser's Forest Flycatcher	R	F	C			GC	f, d, e
Fraseria cinerascens	White-browed Forest Flycatcher	U					GC	w
Melaenornis annamarulae	Nimba Flycatcher			F	VU	UG	GC	d
Muscicapa cassini	Cassin's Flycatcher	C	U				GC	w
Muscicapa olivascens	Olivaceous Flycatcher			R			GC	e
Muscicapa comitata	Dusky-blue Flycatcher	R		R				e
Muscicapa ussheri	Ussher's Flycatcher		U	F			GC	d, e
Myioparus griseigularis	Grey-throated Flycatcher		R	R			GC	e
Myioparus plumbeus	Lead-coloured Flycatcher	R	R					e, o
MONARCHIDAE								
Erythrocercus mccallii	Chestnut-capped Flycatcher		U	U			GC	f, d
Elminia nigromitrata	Dusky Crested Flycatcher	R					GC	f
Trochocercus nitens	Blue-headed Crested Flycatcher	C	F	C			GC	f, d, e
Terpsiphone rufiventer	Red-bellied Paradise Flycatcher	C	F	C			GC	f, d, e
PLATYSTEIRIDAE								
Megabyas flammulatus	Shrike Flycatcher	R	R	R			GC	f, d
Dyaphorophyia castanea	Chestnut Wattle-eye	F	F	F / b			GC	f, d
Dyaphorophyia blissetti	Red-cheeked Wattle-eye	U		R			GC	e
Dyaphorophyia concreta	Yellow-bellied Wattle-eye	R	R					f
Batis poensis	Bioko Batis		R / b				GC	f
PICATHARTIDAE								
Picathartes gymnocephalus	Yellow-headed Picathartes	U			VU	UG	GC	f
TIMALIIDAE								
Illadopsis rufipennis	Pale-breasted Illadopsis	F		F				f
Illadopsis fulvescens	Brown Illadopsis	F		C			GC	e
Illadopsis cleaveri	Blackcap Illadopsis	U		C			GC	f, d
Illadopsis rufescens	Rufous-winged Illadopsis	F	R	C	NT	UG	GC	f, d

continued

Taxa	Common Name	North Lorma	Gola	Grebo	Threat Status	Endemism	GC Forests Biome	Habitat
REMIZIDAE								
Pholidornis rushiae	Tit-hylia		R	U			GC	d, e
NECTARINIIDAE								
Anthreptes gabonicus	Brown Sunbird	R					GC	w
Anthreptes rectirostris	Green Sunbird		F	F			GC	d, e
Anthreptes seimundi	Little Green Sunbird	R		?			GC	f
Deleornis fraseri	Fraser's Sunbird	F	F	C / b			GC	f, d
Cyanomitra cyanolaema	Blue-throated Brown Sunbird	U	C	C			GC	f, d, e
Cyanomitra olivacea	Olive Sunbird	C	C	C				f, d, e
Chalcomitra adelberti	Buff-throated Sunbird		R	R			GC	d, e
Hedydipna collaris	Collared Sunbird	U	C	C				d, e, o
Cinnyris chloropygius	Olive-bellied Sunbird	F	F					o
Cinnyris johannae	Johanna's Sunbird		C	C			GC	f, d, e
Cinnyris superbus	Superb Sunbird		R				GC	d
MALACONOTIDAE								
Malaconotus lagdeni	Lagden's Bush-shrike			R	NT			d, e
Malaconotus multicolor	Many-coloured Bush-shrike	U	F	U				f, d
Dryoscopus sabini	Sabine's Puffback		F	C			GC	f, d
Laniarius leucorhynchus	Sooty Boubou	U					GC	e
PRIONOPIDAE								
Prionops caniceps	Red-billed Helmet-shrike		R	U			GC	f, d
ORIOLIDAE								
Oriolus brachyrhynchus	Western Black-headed Oriole	C	C	C / b			GC	f, d
DICRURIDAE								
Dicrurus atripennis	Shining Drongo	F	F	F			GC	f, d
Dicrurus modestus	Velvet-mantled Drongo	F	R	C				f, d, e
STURNIDAE								
Onychognathus fulgidus	Forest Chestnut-winged Starling	R		R			GC	f, d
Lamprotornis cupreocauda	Copper-tailed Glossy Starling	U	C	C	NT	UG	GC	f, d, e
Cinnyricinclus leucogaster	Violet-backed Starling		R					o
PLOCEIDAE								
Malimbus ballmanni	Gola Malimbe		F / b		EN	UG	GC	f
Malimbus scutatus	Red-vented Malimbe	F	U	F			GC	f, d, e
Malimbus malimbicus	Crested Malimbe	U	U	U			GC	f, d, e
Malimbus nitens	Blue-billed Malimbe	F	U	C / b			GC	f, d, e
Malimbus rubricollis	Red-headed Malimbe			F			GC	d, e
Ploceus nigerrimus	Vieillot's Black Weaver	U		U / b			GC	o
Ploceus cucullatus	Village Weaver	C / b	C / b					o
Ploceus albinucha	Maxwell's Black Weaver		R				GC	f
Euplectes hordeaceus	Black-winged Red Bishop	R						o

continued

Taxa	Common Name	North Lorma	Gola	Grebo	Threat Status	Endemism	GC Forests Biome	Habitat
ESTRILDIDAE								
Nigrita canicapillus	Grey-headed Negrofinch	U	F	C				f, d, e
Nigrita bicolor	Chestnut-breasted Negrofinch	U	F	F			GC	f, d, e
Estrilda melpoda	Orange-cheeked Waxbill	U						o
Spermophaga haematina	Western Bluebill		U / b	R			GC	d, e
Pyrenestes sanguineus	Crimson Seedcracker		C / b				GC	e, o
Spermestes bicolor	Black-and-white Mannikin		F / b	R				o
TOTALS		**143**	**145**	**156**	**14**	**12**	**136**	
			211					

Appendix 8

Gazetteer of localities for small mammal surveys.

Ara Monadjem and Jakob Fahr

Locality	Coordinates	Country
North Lorma National Forest (camp site)	8°01'41"N, 09°43'42"W	Liberia
Gola National Forest (camp site)	7°27'10"N, 10°41'33"W	Liberia
Gola National Forest (S.L.C. village)	7°26'56"N, 10°39'05"W	Liberia
Grebo National Forest (camp site)	5°24'11"N, 07°43'57"W	Liberia
Grebo National Forest (Jalipo village)	5°22'11"N, 07°46'15"W	Liberia
Bomi Wood Concession	c. 07°43'N, 10°29'W	Liberia
Deaple	06°53'N, 08°29'W	Liberia
Du River	c. 06°23'N, 10°22'W	Liberia
Dugbe River (12 mi SSE Jaoudi)	05°27'N, 08°15'W	Liberia
Gabayae (1 km N, 3 km E Zigida)	08°03'N, 09°27'W	Liberia
Harbel	06°16'N, 10°21'W	Liberia
John's Town (near Zozoma, 2 miles SW Voinjama)	08°24'N, 09°46'W	Liberia
Klouga Mtn. (near Voinjama)	08°27'N, 09°43'W	Liberia
Kpeaple	06°36'N, 08°32'W	Liberia
River Peblei (S of Grassfield)	07°29'N, 08°34'W	Liberia
Tars Town (25 km N of Zwedru)	06°13'N, 08°08'W	Liberia
Tokadeh (Nimba Region)	07°27'N, 08°40'W	Liberia
Voinjama	08°25'N, 09°45'W	Liberia
Wologizi Mts. (northern foothills)	c. 08°12'N, 09°52'W	Liberia
Zigida (Wonegizi Mts.)	08°02'N, 09°29'W	Liberia
Zigida (7 km N, 2 km E)	08°06'N, 09°28'W	Liberia
Zigida (7 mi N, 1 mi E)	08°08'N, 09°28'W	Liberia
Zigida (10.5 km N, 1 km E)	08°08'N, 09°28'W	Liberia
Zigida (11 km N, 2 km E)	08°08'N, 09°28'W	Liberia
Zigida (11 km N, 3 km E)	08°08'N, 09°27'W	Liberia
Zigida (13 km N, 1 km E)	08°09'N, 09°28'W	Liberia
Zwedru (Tchien)	06°04'N, 08°08'W	Liberia
Balouma (25 km NW Macenta)	08°39'N, 09°38'W	Guinea
Gola Forest Camp (4 mi S Lalehun)	07°38'N, 10°58'W	Sierra Leone
Forêt Classée du Cavally (33 km W Zagné)	06°10'N, 07°47'W	Ivory Coast
Taï National Park (IET-station)	05°50'N, 07°21'W	Ivory Coast
Baké River Bridge (1 km S, 1.25 km W Baro)	05°16'N, 09°13'E	Cameroon

Appendix 9

Shrews and rodents collected during the RAP survey and deposited in the collections of the Zoologisches Forschungsmuseum Alexander Koenig, Bonn (ZFMK).

Ara Monadjem and Jakob Fahr

Species	Locality	Catalogue N°
Crocidura muricauda	Grebo 1	ZFMK 2006.11
	North Lorma	ZFMK 2006.12
Crocidura obscurior	North Lorma	ZFMK 2006.13
Hylomyscus alleni	Grebo 1	ZFMK 2006.14
	Gola 2	ZFMK 2006.15
	North Lorma	ZFMK 2006.16

Appendix 10

Revised checklist of bat species recorded from Liberia (modified from Koopman et al. 1995).

Jakob Fahr

Habitat: Coarse assignment to preferred habitat type (F: forest; S: savanna and woodland; in parentheses: marginally used habitat).
Red List: Global threat status (IUCN 2006): EN = Endangered; VU = Vulnerable; NT = Near Threatened; LC = Least Concern; n.a.: not assessed.
RAP survey: Species recorded during the survey of North Lorma, Gola and Grebo National Forests, November–December 2005, are marked with 'X'.

Species	Habitat		Red List	RAP survey
Pteropodidae				
Micropteropus pusillus		S	LC	
Epomops buettikoferi [1]	F	(S)	LC	X
Hypsignathus monstrosus	F	(S)	LC	X
Nanonycteris veldkampii	F	(S)	LC	X
Scotonycteris zenkeri	F		NT	X
Scotonycteris ophiodon	F		**EN**	
Megaloglossus woermanni	F		LC	X
Myonycteris torquata	F	(S)	LC	X
Lissonycteris angolensis smithii	F	(S)	LC	
Rousettus aegyptiacus	F	S	LC	X
Eidolon helvum	F	S	LC	
Emballonuridae				
Saccolaimus peli	F		NT	
Nycteridae				
Nycteris intermedia	F		NT	
Nycteris arge	F		LC	X
Nycteris major	F		**VU**	
Nycteris grandis	F	(S)	LC	
Nycteris hispida	F	S	LC	
Nycteris macrotis	F	S	LC	
Rhinolophidae				
Rhinolophus (simulator) alticolus [2]	F		n.a.	
Rhinolophus hillorum	F		**VU**	X

continued

Species	Habitat		Red List	RAP survey
Rhinolophus alcyone	F	(S)	LC	X
Rhinolophus guineensis	F		VU	
Rhinolophus landeri [3]	(F)	S	LC	
Rhinolophus ziama [4]	F		EN	
Hipposideridae				
Hipposideros jonesi	F	S	NT	
Hipposideros marisae	F		EN	
Hipposideros caffer [5]	F	S	LC	
Hipposideros ruber	F	(S)	LC	X
Hipposideros fuliginosus	F		NT	X
Hipposideros beatus	F		LC	X
Hipposideros cyclops	F		LC	X
Hipposideros gigas [6]	F		LC	X
Vespertilionidae				
Kerivoula lanosa muscilla	F	(S)	LC	
Kerivoula phalaena [7]	F		NT	
Myotis tricolor	F	S	LC	
Myotis bocagii	F	S	LC	X
Pipistrellus hesperidus [8]	(F)	S	LC	
Pipistrellus nanulus	F	(S)	LC	
Hypsugo (crassulus) bellieri [9]	F		n.a.	X
Neoromicia nanus	F	S	LC	X
Neoromicia guineensis [10]	(F)	S	LC	X
Neoromicia capensis	(F)	S	LC	
Neoromicia aff. *grandidieri* [11]	F		n.a.	X
Neoromicia brunneus	F		NT	
Neoromicia tenuipinnis	F	(S)	LC	X
Neoromicia rendalli [12]	(F)	S	LC	
Mimetillus m. moloneyi	F	(S)	LC	
Glauconycteris poensis	F		LC	X
Scotophilus nux [13]	F		LC	
Miniopterus villiersi [14]	F	(S)	n.a.	
Miniopterus inflatus	F		LC	
Molossidae				
Chaerephon b. bemmeleni	F		LC	
Chaerephon major	(F)	S	LC	
Chaerephon pumilus	(F)	S	LC	
Mops nanulus	F	(S)	LC	
Mops spurrelli	F		LC	

continued

Species	Habitat		Red List	RAP survey
Mops (brachypterus) leonis [15]	F		LC	
Mops thersites	F		LC	
Mops condylurus	(F)	S	LC	

Remarks:

1) Kuhn (1965) mentions *Epomophorus gambianus* from Kpeaple. However, Bergmans (1988) and Koopman et al. (1995) question the species identification; it is likely that it represents *Epomops buettikoferi.*

2) The taxon *alticolus*, known from the highlands in the forest zone of Liberia, Guinea and Cameroon, and from the Jos Plateau of Nigeria, probably represents a species distinct from *R. simulator* (Csorba et al. 2003).

3) An unpublished specimen of *R. landeri* from Zwedru (SMNS 38565) represents the first record for Liberia.

4) Koopman et al. (1995) published a record of *R. maclaudi* from NW Liberia, but the specimen was later assigned to a new species, *R. ziama* (Fahr et al. 2002).

5) *Hipposideros lamottei*, described from the Guinean side of Mt. Nimba, was reported from Liberia and other West African countries by Koopman et al. (1995). Specimens were re-examined and found to represent *H. caffer* (Decher and Fahr 2007).

6) The forest population of what was formerly called *H. commersoni* on mainland Africa is now referred to *H. gigas* (Simmons 2005, Decher and Fahr 2007).

7) Happold (1987) and Koopman et al. (1995) erroneously listed *K. smithii* for Liberia and Côte d'Ivoire (Fahr in press-b).

8) Liberian records of *P. kuhlii* and *P. rusticus* (Hill 1982, Wolton et al. 1982, Koopman et al. 1995) are here tentatively referred to *P. hesperidus* (see Kock 2001).

9) The taxon *bellieri*, endemic to Upper Guinea, is probably specifically distinct from *Hypsugo crassulus* (Fahr in press-c).

10) Kuhn (1962, 1965) published a record of *Eptesicus minutus* from Deaple, which was tentatively referred to *Pipistrellus somalicus* by Koopman et al. (1995); based on published measurements, the specimen is more likely to represent *Neoromicia guineensis.*

11) West African specimens from Côte d'Ivoire (J. Fahr unpubl.) and Liberia (this study) belong either to *N. grandidieri* sensu Thorn et al. (in press) or represent an undescribed species.

12) The single Liberian record of *N. rendalli* from Harbel by Kuhn (1965) is questionable as this is a savanna species; the specimen should be re-examined.

13) Following the taxonomy of Robbins et al. (1985), Liberian records variably called *S. dinganii* (Hill 1982, Wolton et al. 1982) or *S. leucogaster nux* (Koopman 1989) are referable to *S. nux.*

14) Specimens from northwestern Liberia published as *M. schreibersii villiersi* (Koopman et al. 1995) represent a distinct species, *M. villiersi* (Fahr et al. 2006).

15) Though currently considered a subspecies of *Mops brachypterus*, the taxon *leonis* is probably specifically distinct from the latter (Beaucournu and Fahr 2003).

Appendix 11

Additional bat species previously recorded from localities near RAP survey sites.

Jakob Fahr

Species	Locality	Reference / Specimen[1]
Gola		
Myotis tricolor	Bomi Wood Concession	Koopman 1989; AMNH 257053
Mimetillus moloneyi	Gola Forest Camp (4 mi S Lalehun), SL	USNM 545601–545603
Mops spurrelli	Gola Forest Camp (4 mi S Lalehun), SL	Grubb et al. 1999; USNM 545613
Mops thersites	Gola Forest Camp (4 mi S Lalehun), SL	USNM 545614–545622
Mops condylurus	Gola Forest Camp (4 mi S Lalehun), SL	USNM 545604–545612
North Lorma[2]		
? Micropteropus pusillus	Zigida	USNM 541527; specimen not examined, possibly *Nanonycteris veldkampii*
Epomops buettikoferi	Zigida; Zigida (10.5 km N, 1 km E; 11 km N, 3 km E)	AMNH 265663, 265664; USNM 541518–541526
Hypsignathus monstrosus	Zigida (7 km N, 2 km E)	AMNH 265661, 265662
Scotonycteris zenkeri	Zigida (11 km N, 3 km E)	AMNH 265675–265678
Megaloglossus woermanni	Zigida; Zigida (10.5 km N, 1 km E; 11 km N, 2 km E; 13 km N, 1 km E)	AMNH 265695–265706; USNM 541528–541531
Lissonycteris angolensis	Zigida; Zigida (10.5 km N, 1 km E)	AMNH 265639–265648; USNM 541510
Nycteris intermedia	Zigida (13 km N, 1 km E)	AMNH 265707
Nycteris grandis	Zigida (7 miles N, 1 mile E)	AMNH 265825
Rhinolophus simulator	Zigida (10.5 km N, 1 km E)	AMNH 265746, 265747
Rhinolophus hillorum	Zigida (10.5 km N, 1 km E); Wologizi Mts. (northern foothills)	AMNH 265709, 265710; SMNS 39671
Rhinolophus guineensis	Zigida (10.5 km N, 1 km E; 11 km N, 3 km E; 13 km N, 1 km E)	AMNH 265711–265745, 265831, 265832, 265919
Rhinolophus ziama	Zigida (7 mi N, 1 mi E)	Koopman et al. 1995 as *R. maclaudi*; AMNH 265708
Hipposideros jonesi	Zigida (13 km N, 1 km E)	AMNH 265749
Hipposideros marisae	Zigida (13 km N, 1 km E)	AMNH 265750
Hipposideros gigas	Zigida (13 km N, 1 km E)	Koopman et al. 1995 as *H. commersoni*; AMNH 265748
Neoromicia nanus	Zigida	USNM 541532
Neoromicia tenuipinnis	Zigida; Zigida (7 mi N, 1 mi E)	AMNH 265826; USNM 541533, 541534
Miniopterus villiersi	Zigida (10.5 km N, 1 km E; 13 km N, 1 km E)	Koopman et al. 1995 as *M. schreibersi*; AMNH 265828–265830
Miniopterus inflatus	Gabayae (1 km N, 3 km E Zigida)	AMNH 265827

[1] Specimens from the American Museum of Natural History, New York (AMNH), the Staatliches Museum für Naturkunde Stuttgart (SMNS), and the National Museum of Natural History, Smithsonian Institution, Washington, DC (USNM).

[2] Most records published by Koopman et al. (1995) as from "Wonegizi Mts. near Ziggida".

Appendix 12

Large mammal species recorded in North Lorma, Gola and Grebo National Forests.

Abdulai Barrie

Site: NL=North Lorma, Go=Gola, Gr=Grebo
Evidence: H=heard, S=Seen, T=Tracks, P=Phototrap, O=Other
(#)=number of individuals, (*)=heard >20 times
IUCN Status (2006): EN=Endangered, VU=Vulnerable, LR/cd=Lower Risk/conservation dependant, LR/nt=Lower Risk/near threatened, LR/lc=Lower Risk/least concern, LC=Least Concern

Species	Common Name	NL	Go	Gr	Evidence	IUCN
PRIMATES						
Hominidae						
Pan troglodytes verus	Chimpanzee	x		x	H[Gr (3)] O[NL (nests), Gr (nut cracking, dung)]	EN
Colobidae						
Piliocolobus badius	Western Red Colobus	x		x	S[NL (3), Gr (25+)]	EN
Colobus polykomos	Western Pied Colobus	x		x	H[Gr (2)] S[Gr (3)]	LR/nt
Procolobus verus	Olive Colobus			x	S[(5+)]	LR/nt
Cercopithecidae						
Cercocebus atys	Sooty Mangabey	x	x	x	H[Go (2*), Gr (4)] S[Go (10+), Gr (12+)] O[Go (2 seen in village)]	LR/nt
Cercopithecus diana	Diana Monkey	x		x	H[NL (2), Gr (*)] S[NL (1), Gr (50+)]	EN
Cercopithecus campbelli	Campbell's Monkey	x	x	x	H[NL (2), Go (6*), Gr (*)] S[NL (2), Gr (15+)]	LR/lc
Cercopithecus petaurista	Lesser Spot-nosed Monkey	x		x	H[Gr (*)] S[NL (2), Gr (13)]	LR/lc
Galagonidae						
Galagoides demidoff	Demidoff's Galago	x	x	x	H[NL (3), Gr (*)] S[Gr (7)]	LR/lc
CARNIVORA						
Felidae						
Panthera pardus	Leopard			x	T O[dung]	LC
Herpestidae						
Herpestes sanguinea	Slender Mongoose	x	x	x	S[NL (1), Go (2), Gr (1)]	LR/lc
Atilax paludinosus	Marsh Mongoose	x	x	x	S[NL (5), Go (8), Gr (2)] T[Gr] O[Go (1 seen in village)]	LR/lc
Viverridae						
Civettictis civetta	African Civet	x		x	T[Gr] O[NL (dung), Gr (dung)]	LR/lc
Nandinidae						
Nandinia binotata	African Palm Civet	x	x	x	H[NL (*), Gr (*)]	LR/lc

continued

Species	Common Name	NL	Go	Gr	Evidence	IUCN
HYRACOIDEA						
Procaviidae						
Dendrohyrax dorsalis	Western Tree Hyrax	x	x	x	H[NL (5), Gr (*)]	LC
PROBOSCIDEA						
Elephantidae						
Loxodonta africana cyclotis	Forest Elephant	x	x	x	T[NL, Go, Gr] O[NL (dung), Go (dung), Gr (dung)]	VU
ARTIODACTYLA						
Bovidae						
Cephalophus dorsalis	Bay Duiker	x	x	x	S[NL (1), Go (1), Gr (1)] T[Go, Gr]	LR/nt
Cephalophus jentinki	Jentink's Duiker			x	P[(1)]	VU
Cephalophus maxwelli	Maxwell's Duiker	x	x	x	S[NL (1), Go (1), Gr (3)] T[NL, Gr] P[NL (1), Gr (2)] O[Gr (dung)]	LR/nt
Cephalophus niger	Black Duiker	x	x	x	S[NL (1), Gr (1)] T[Go, Gr]	LR/nt
Cephalophus ogilbyi	Ogilby's Duiker	x		x	S[NL (1), Gr (1)] T[NL] O[NL (dung)]	LR/nt
Cephalophus silvicultor	Yellow-backed Duiker			x	T P[(1)]	LR/nt
Syncerus caffer	African Buffalo	x			T O[dung]	LR/cd
Tragelaphus euryceros	Bongo			x	T	LR/nt
Tragelaphus scriptus	Bushbuck	x	x	x	S[Gr (1)] T[NL, Go, Gr] O[Go (dung), Gr (dung)]	LR/lc
Hippopotamidae						
Hexaprotodon liberiensis	Pygmy Hippopotamus			x	T O[dung]	EN
Suidae						
Potamochoerus porcus	Red River Hog			x	T O[rooting]	LR/lc
RODENTIA						
Hystricidae						
Atherurus africanus	Brush-tailed Porcupine	x	x	x	T[Go, Gr]	LC
PHOLIDOTA						
Manidae						
Uromanis tetradactyla	Long-tailed Pangolin		x	x	O[Go (scales), Gr (scales & feeding site)]	LR/lc

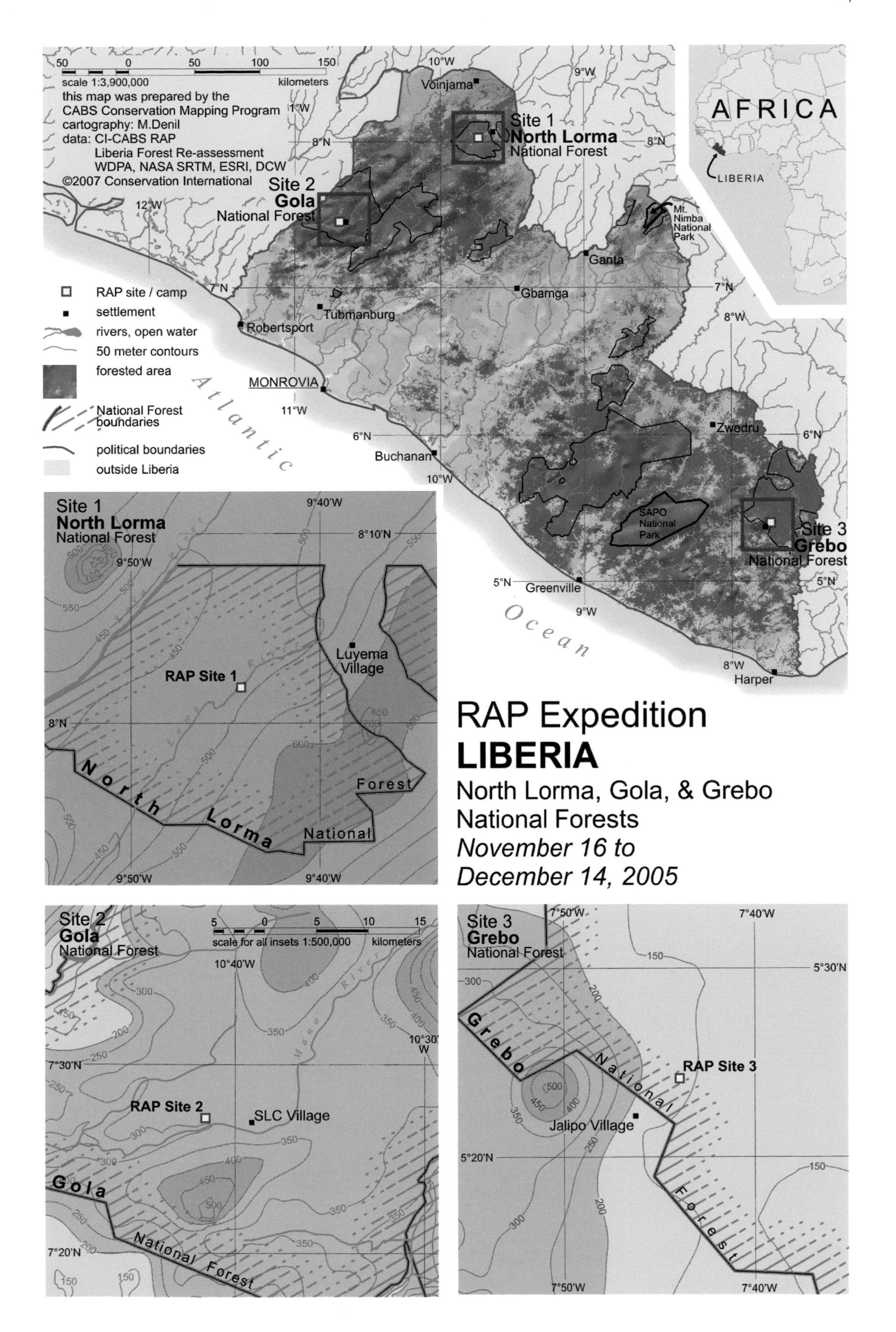

RAP Expedition
LIBERIA

North Lorma, Gola, & Grebo National Forests

November 16 to December 14, 2005

Annika Hillers

The many rocky streams in Gola National Forest represent a typical habitat for a number of frogs including the Endangered *Amnirana occidentalis*.

Peter Hoke

The community of Luyema welcoming the RAP team.

Annika Hillers

Bufo togoensis, a Near Threatened forest species, was recorded in both North Lorma and Grebo National Forests.

Peter Hoke

Ara Monadjem and local guide setting up mist nets to capture bats in North Lorma National Forest.

Peter Hoke

Aerial view of forest in northwestern Lofa County.

Peter Hoke

A large rock with 20 Yellow-headed Picathartes, *Picathartes gymnocephalus*, nests in good condition in North Lorma Nationa Forest. This Upper Guinea endemic is a generally scarce and very local resident in the forest zone. Liberia probably holds the largest population of this Vulnerable bird species.

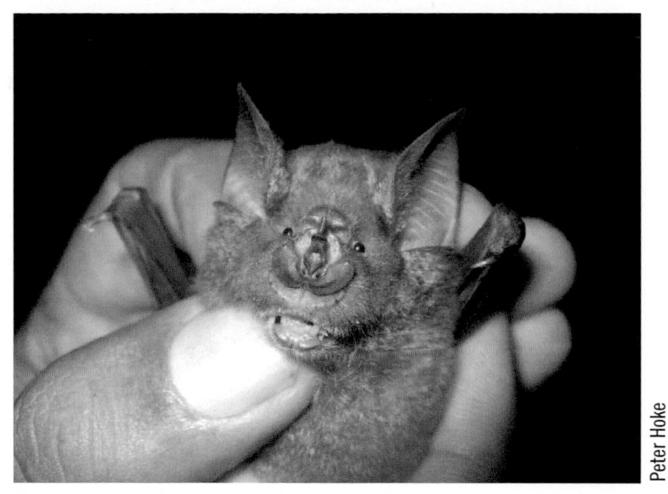

Peter Hoke

The records of *Rhinolophus hillorum* from Gola National Forest constitute a range extension of approximately 100 km to the southwest. It is a near-endemic to West Africa and is listed by IUCN as Vulnerable due to habitat loss within its limited distribution.

Peter Hoke

Members of the RAP team at the UNMIL compound in Fishtown.

Peter Hoke

Amandu K. Daniels (left) and Carel Jongkind (right) noting *Drypetes* sp., a plant species new to science, in Grebo National Forest.

Annika Hillers

The Endangered *Phyronbatrachus* cf. *annulatus* found in Grebo National Forest. Further genetic analyses will clarify if this specimen can be referred to a known species (*Phrynobatrachus annulatus*) or if it is new to science and thus probably a Liberian endemic.

Peter Hoke

A scaly-tailed squirrel, *Anomalurus* cf. *pusillus*, found inside a tree in Grebo National Forest. This is only the third record of this species for West Africa.

Carel C.H. Jongkind

River next to the base camp in North Lorma National Forest.

Carel C.H. Jongkind

Cola buntingii, an Upper Guinea endemic plant species.

Peter Hoke

Unloading gear from an UNMIL helicopter to pickup trucks at Voinjama.

Carel C.H. Jongkind

Psychotria ombrophila, an Upper Guinea endemic plant species.

Ara Monadjem

White-browed Forest Flycatcher, *Fraseria cinerascens*, an Upper Guinea endemic, that was mist-netted in North Lorma National Forest.